A Review of Research in Mathematical Education

Prepared for the Committee of Inquiry into the
Teaching of Mathematics in Schools

Part A   Research on Learning and Teaching

A.W. Bell, J. Costello and D.E. Küchemann

NFER-NELSON

Published by The NFER-NELSON Publishing Company Ltd.,
Darville House, 2 Oxford Road East,
Windsor, Berks. SL4 1DF

First Published 1983
© A.W. Bell, D. Küchemann and J. Costello, 1983
ISBN 0-7005-0612-8
Code 8144 02 1

Printed in England

Distributed in the USA by Humanities Press Inc.,
Atlantic Highlands, New Jersey 07716 USA

A Review of Research in Mathematical Education

Part A

HF

Contents

List of Abbreviations of Journal Titles

Full Title	Abbreviation
Acta Psychologica	Acta Psych
American Educational Research Journal	Am Ed Res J
American Psychologist	Am Psych
Arithmetic Teacher (USA)	Arith Teacher
Australian Journal of Education	Aust J Ed
British Journal of Educational Psychology	B J Ed Psych
British Journal of Psychology	B J Psych
British Journal of the Society for Clinical Psychology	B J Soc Clinical Psych
California Journal of Educational Research (USA)	Calif J Ed Res
Canadian Journal of Behavioural Science	Canadian J Beh Sci
Canadian Psychological Review	Canadian Psych Rev
Child Development	Child Dev
Cognitive Psychology	Cog Psych
Cognitive Science	Cog Sci
Developmental Psychology	Dev Psych
Dissertation Abstracts International (USA)	Diss Abs Int
Education in Chemistry	Ed Chem

Education for Teaching	Ed for Teaching
Education Research	Ed Res
Educational Studies	Ed Stud
Educational Studies in Mathematics	Ed Stud Math
Educational Technology	Ed Tech
European Journal of Science Education	Eur J Sci Ed
Human Development	Human Dev
International Journal of Mathematical Education in Science and Technology	Int J Math Ed Sci Tech
Journal of the British Association of Teachers of the Deaf	J Br Ass Teachers of the Deaf
Journal of Children's Mathematical Behaviour (USA)	J Child Math Beh
Journal of Curriculum Studies	J Curr Stud
Journal of Educational Psychology	J Ed Psych
Journal of Educational Research	J Ed Res
Journal of Experimental Child Psychology	J Exp Child Psych
Journal of Experimental Education	J Exp Ed
Journal of Experimental Psychology (USA)	J Exp Psych
Journal of General Psychology	J Gen Psych
Journal of Occupational Psychology	J Occup Psych
Journal of Psychology	J Psych
Journal for Research in Mathematical Education (USA)	J Res Math Ed
Journal of Research in Science Teaching	J Res Sci Teaching
Journal of Social Issues	J Soc Issues

Journal of Verbal Learning and Verbal Behaviour	J Verbal Learning and Verbal Behaviour
Mathematics in School	Maths in School
Mathematics Teacher (SA)	Maths Teacher
Mathematics Teaching (Br)	Maths Teaching
Monograph of the Society for Research in Child Development	Monog Soc Res Child Dev
National Elementary Principal	Nat El Principal
Physics Education	Phys Ed
Programmed Learning and Educational Technology	Programmed Learning and Ed Technol
Psychological Bulletin	Psych Bull
Psychological Monograph	Psych Monog
Psychological Review	Psych Rev
Quarterly Journal of Experimental Psychology	Quarterly J Exp Psych
Research in Education	Res Ed
Review of Educational Research	Rev Ed Res
School Science Review (Br)	Sch Sci Rev
School Science and Mathematics (USA)	Sch Sci Math
Studies in Science Education	Studies in Science Ed
Times Educational Supplement	Times Ed Supp

Preface

This review was prepared for the Cockcroft Committee of Inquiry into the Teaching of Mathematics in Schools, and its main conclusions were incorporated in their report, *Mathematics Counts* (HMSO, 1982). Its aim was to display the main outcomes of research relevant to the teaching and learning of mathematics. For this publication, a supplement has been added to take account of work published since the original review was compiled.

We acknowledge with gratitude the help of members of the Consultative Group: Dr M. Brown, Mrs M.R. Eagle, Mr D.S. Fielker, Mrs A. Floyd, Dr G.B. Greer, Dr K. Hart, Dr A.G. Howson, Professor D.C. Johnson, Professor K. Lovell, Professor R.R. Skemp and Mr J.D. Williams.

We are also indebted, for particular contributions, to Professor Hugh Burkhardt, Dr Brian Greer, Dr John Head and Professor D.C. Johnson.

A special acknowledgment is due to Marian Martin, whose contribution has included responsibility for the typing and much of the administrative detail concerned in the production of this review.

Introduction

1. CURRENT PROBLEMS

In this introduction we intend to highlight some of the main
points of this review of research in mathematical education,
and at the same time to put it into the context of the
questions currently under debate by the Committee, the profession and the public. We consider first what research has
to offer towards the explanation and interpretation of those
manifestations of anxiety which brought the Committee into
being. Today's *Times Educational Supplement* provides a
typical headline, 'Low Scores in Maths Test'; this covers a
short paragraph of statements such as 'sixty per cent of the
pupils could not write 4.867 to two decimal places'. These
anxieties and complaints probably owe as much to social and
cultural factors as they do to the realities of mathematical
attainment; the pendulum has swung in the last ten years
away from easier divorce and abortion and towards higher pay
for the police. It is also part of a world-wide trend;

1

nearly all countries are in a period of cutbacks in development activities. In France, the IREMs have their budgets
slashed; in the US, money for the implementation of even
highly successful development projects has been cut off.
(Dr. Howson documents the historical and international comparisons in Section C of this review.) Perhaps the most
depressing aspect of this atmosphere of complaint is the
willingness of the old to castigate the young - or, more
particularly, their teachers - for inability to answer questions which they themselves cannot answer, and for which they
have no great use. As one lady on a phone-in said, 'We were
taught properly when I was at school. I know that point
nought seven is seven tenths'.

2.   IS MATHEMATICAL ATTAINMENT LOW?

But what can be said about the specific complaints? Are
these scores indeed 'low'? If so, why are they, and what
actions are likely to improve them?

    First, has attainment declined over time? The evidence
currently available suggests that it has not. Submissions
received by the Expenditure sub-committee claimed that test
results of apprentices entering certain employments (e.g.,
the Navy) had declined over a decade or so; but at the same
time pointed out that the stratum of ability from which these
entrants were drawn was lower (Report 1966-7, pp.764-5).
The only hard evidence *published* was from Coventry (Gilbert,
1975, in the *Times Educational Supplement*). This appeared
to show that, in the group of entrants tested, Reasoning and
Mechanical Aptitude (the presumed basic abilities) had
declined only very slightly, English and Arithmetic scores had
declined substantially. However, the continuation of the
same research for the years 1975-78 showed results back at
1970-71 levels, and the total pattern of results for the nine
years now looks like random fluctuations than even a U-shaped
trend (T.E.S., 23rd May 1980).

    Longer-term comparisons (e.g., McIntosh, 1977) make it
clear that dissatisfaction with the mathematical understanding
of the young has a long history. It is one of the easiest
weapons with which to assert superiority. More recent public
comments (e.g., IMA submission, 1980) tend expressly to
avoid allegations of decline, asserting rather that current
attainments are inadequate in relation to the need.

    Secondly, how does British attainment compare with that
in other countries? This is hard to determine. In Chapter
Twelve we have extracted approximately comparable items from
British and American national surveys. No items are identical, and American testing is at ages nine, 13, 17 and adult,
whereas ours is at 11 and 15. All one can conclude is that
the general pattern of results is the same, i.e., fair
performance on routine calculation, much lower on applications

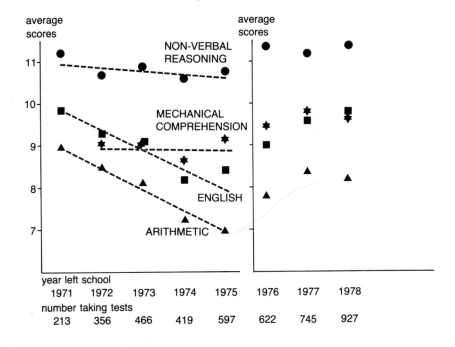

Left: How standards appeared to fall between 1971 and 1975 and,
right, the subsequent upswing.

to fresh situations and on items containing some unusual
aspect. A more obvious difference is in the range of
curriculum available for testing across the general school
population in the US; in algebra, geometry and statistics
this is more limited than in the British tests. The few
scraps of comparable information in Swedish and German tests
bear out the same conclusions – some differences in curriculum emphasis but broadly similar patterns of result.

Are current attainments 'inadequate'? Further light
should be thrown on this question by the other current
DES/Committee of Inquiry research projects. Whatever the
detailed outcomes of these, they are likely to show that
industrial and commercial practices become adapted to conform
to the capacities of the workers involved. For example,
since proportion is not generally well understood, nurses
are in certain cases given detailed specific direction about
how to make up a solution of the correct strengths using
marked pictures of a syringe; as another example, no bank
would rely on the calculating accuracy of its clerks un-
assisted by machines. Some suggestions of priorities to
be attached to numerical as opposed to geometrical work may
emerge. But nothing is likely to challenge the general
conclusion that any obtainable increase in genuine permanent

mathematical understanding would represent a valuable increase
in personal potential, both in employment and in personal
life.

A more difficult question may be what aspects of mathematical attainment should have priority in schools, and
which are better learned 'on the job'.

3. THE DIFFICULTY OF LEARNING MATHEMATICS

Is there any intrinsic reason why mathematical understanding
is so hard to achieve? Is it reasonable to assume that harder
work by pupils and by teachers could achieve sufficiently good
results to satisfy the complainants? The answer to the first
question is yes, plenty, and therefore to the second is in
general no, and we shall explain subsequently what kinds of
better-directed work may be helpful.

Briefly, mathematics is hard because it deals with
abstract relationships, but some parts of it look easy because
they consist of symbol-manipulations, which it is possible to
learn without understanding the underlying operations to which
they refer. The justification for these statements lies in
the chapters which follow, but it may be helpful to discuss
them briefly now. Consider the following question: Minced
beef costs 88.2 pence per kilogram; I buy 0.58 kg; to find the
cost, which calculation must I do: (A) $0.58 \div 88.2$,
(B) $88.2 \div 0.58$, (C) 0.58×88.2 or (D) $88.2 - 0.58$? The
reader who wishes to gain insight should give himself five
seconds to jot down the answer A, B, C or D starting NOW .....
Having done that, consider whether you are sure of the answer;
whether you had to pause to think for a moment; and whether
the time limit interfered with your thinking. (You may have
experienced a case of the Yerkes-Dodson law, vividly described
by Skemp, as follows. If you are in a field and are being
chased by a bull you will probably perform above your usual
level at jumping the gate, but below usual at undoing the
combination lock. Anxiety may enhance motor skills, but it
depresses complex intellectual performances.) The question
above was correctly answered by 30 per cent of a representative sample of fourth year secondary pupils; it illustrates
the importance of distinguishing between the ability to
calculate (with decimals, in this case) and the ability to
choose the correct operation. It also underlines the fact
that the second of these is sometimes very hard - and it
cannot, as the calculation can, be performed by a machine. It
is also the case that the *choice of operation* - not just the
calculation - is much easier if the figures are small whole
numbers. This is an example of an aspect of arithmetic to
which little attention has been given in the past - because
we failed to realise that it was a serious learning difficulty.

4

As a second example which may help in seeing some of the less obvious difficulties in mathematics, consider the now notorious question of Boycott's batting average - 500 runs, 5 times out. How would you explain to a young boy what his average would have been if he had scored 532 runs, and been 5 times out? 'You divide the 532 by 5. That gives 106.4 ...' 'Yes, but what does his average *mean*?' 'Well, it's what he would have scored in each innings if he had scored the same each time ...' 'But how could he have scored 106.4?...' Can you continue convincingly?

Chapter Six describes research which identifies levels of difficulty in a wide range of material from the mathematics curriculum. Most people will find here plenty of surprises: aspects of the subject which present unexpected difficulty. This should serve both to show how complex a task it is to manage the successful learning of mathematics, and also to provide the teacher with material to enlighten him and help him in this task.

The particular problem of keeping the connection between the symbolic manipulations and their meaning is discussed in Chapter Five, and frequently elsewhere; that of choice of operation and performance of the calculation in Chapter Four. A possibly different reason for the difficulty of learning mathematics is treated in Chapter Three; this is that some parts of mathematics may require modes of logical thinking which some children do not develop until late adolescence, and some not at all (Piaget's formal reasoning stage).

4. TEACHING METHODS

If it is accepted that teaching mathematics is difficult, what has research to say about what methods are likely to be more successful? To make any progress in this question it is necessary first to distinguish between Skills, Conceptual Structures and General Strategies. This is the subject of Chapter Four. Briefly, conceptual structures are richly interconnected bodies of knowledge (including 'programs' governing skills), and it is these which are the substance of mathematical knowledge stored in the memory. Skills needed at high level of fluency, and facts required at immediate recall need to be not only understood and embedded in the conceptual structure, but also brought up to the level of immediacy or fluency by regular practice. General Strategies include the ability to approach a problem with some confidence that experiment can lead to a solution, and that a first result can usually be checked by an argument or by some other method. With these we should include general appreciation of what mathematics is and what it can achieve. The orientation given to the work by the teacher has been shown to govern the kind of knowledge which the pupils build up, determining, for example, whether it is superficial and rote-memorized, or understood and deep. The recognition that these three elements, Skills and Facts, Conceptual Structures, and General Strategies and Appreciation involve distinct aspects of teaching and require separate attention pervades the research on teaching. The development of teaching which included all three aspects would represent a useful advance; at present many teachers tend to pursue one to the exclusion of the others.

The chapter divisions reflect these divisions: Chapters Six and Seven on Teaching Conceptual Structures, Chapter Six focusing on research which builds 'conceptual maps' of the subject matter, while Chapter Seven considers aspects of teaching method; then Chapter Eight discusses both the levels of difficulty and experiments on teaching of General Strategies, or Process Aspects of Mathematics. Chapter Nine is more specifically about the development of attitudes to mathematics. The subsequent chapter titles are self-explanatory.

After the distinction between the skills, concepts and strategies, the next most powerful results concerning teaching method take us into the question of what constitutes effective, meaningful teaching - in general, and in particular topics. The most successful methods, and the most plausible theory, use 'cognitive conflict' rather than the accumulation of skills (Chapter Seven, sections 5, 6). This method presents a problem, allows the learner to use his natural approach, but arranges for him to reach a contradiction. He is then shown the new approach, which enables him to resolve the contradiction. The importance of immediate feedback of correctness, and the requirement of mastery of one (small) aspect before progressing to another, are also shown (Chapter Seven, section 10). In general, these methods might be termed Diagnostic Teaching, the greatest departure from current general practice would be in the closer monitoring by the teacher of the *conceptual understanding* of the learner - diagnosing the nature of misunderstandings more deeply than normal marking does - and the design of problems *for discussion* which enable the conceptual errors to be exposed and thence corrected by the pupil.

5. CONTENT OF THE CURRICULUM

What becomes clear from the material reported in Chapter Six, and in particular from the CSMS and the APU surveys, is that there is a very great distance between what can be achieved by the most and the least able pupils, with understanding - far greater than has yet been recognized in syllabuses. (This makes the proposals for a three-paper model, 1 + 2 or 2 + 3, for the 16+ examinations look highly dubious; an approach in which some 12 - 15 modules were offered, from which about five were chosen by the candidate, would offer greater possibility for meeting the needs and less rigid labelling of pupils by levels.) It is the curriculum for the least able pupils which presents the most acute difficulties. A common practice is to restrict their work to computation. This is almost certainly the reverse of the best policy. The research shows (1) that if computational skills are not underpinned by conceptual understanding they are not easily retained, and also that disastrous mistakes occur (e.g. Chapter Five, section 5); (2) that choosing the operation can itself be very hard (Chapter Four, section 2); and (3) that meaningful methods are relatively more successful with the less able than the more able (Chapter Seven, section 6). It is also known that transfer of principles from one context to another is unexpectedly hard - in fact, that what to a mathematician looks like the same principle may be a very different procedure in a different context (see Chapter Two, section 3). (An illustration of a similar situation occurred when some 14-year-olds were asked to find the cost of a litre of petrol, 0.22 gal, if (when!) a gallon cost £1.20. Encouraged to try easier numbers, they said 'About 1/4 gal, so 30 pence'. What to do with the 0.22 then? 'Divide, since we divided to get 30 pence'. A better, if less vivid illustration occurred when pupils, who had worked a sequence of proportion questions concerning the strength of orange juice (e.g. 3 glasses of orange and 5 water in one jug, 9 orange and ? water to give the same strength in another jug), went on to mathematically identical problems on the sizes of a man and his son in two photographs of different sizes. The tendency to equate differences rather than ratios was much stronger in this case, in fact some pupils were quite convinced that this was the correct method for photographs.)

It follows that the less able need to spend considerably more time on problems involving the application of a limited range of mathematical principles to a greater variety of different contexts. This could form an interesting and valuable course. Numbers could not entirely be kept simple; the research shows that, for example, proportion problems in which the ratios consist of doubling and trebling are quite different - and are often done by different methods - from those with harder ratios. Real problems with realistic numbers are needed; this means it is essential to use a calculator. In any case, with pupils of limited learning capacity, if the necessary time is spent to teach them 'long' computations, they will lose the time they need for learning how to apply the mathematics, which no machine can do for them. (They also need to be able to round the numbers and check the size of their answer mentally.)

In general the guidance on the question of curriculum content given by the research is that coherent bodies of knowledge should be chosen, within which pupils concerned can achieve a high degree of mastery. The low retention of fragmentary knowledge is well attested: also clear is the degree of intensity needed to achieve long-term retention and transfer to new situations which is shown in the successful teaching experiments (Chapter Seven, especially sections 5, 6).

6. EVALUATION

Chapter Twelve discusses both the *results* of some national and international evaluations, and also some of the newer modes of evaluation which have been developed, mainly in response to the realization that there are many more facets of understanding to explore than are revealed by traditional test material. The breadth of material developed by the APU, demonstrably necessary to obtain even a reasonably adequate sampling of the curriculum, should make it clear to local authorities that more limited forms of test cannot possibly give satisfactory information on the performance either of pupils or of schools, and are liable to produce gross distortions of the curriculum. The demonstration that inter-view testing can provide both satisfactory and highly valuable material also represents a substantial advance.

7. CALCULATORS AND COMPUTERS (Chapter Thirteen)

By now a considerable number of research studies in the United States have compared the number performance of pupils who have used calculators with that of groups who have not. The weight of evidence is strong in showing that the cal-cular groups perform at least as well as the others. This research needs to be given greater publicity. We do not wish to have to repeat it in Britain. Our rather small-scale British projects are preparing material to help in using the calculator as an aid in the teaching of the normal mathematics curriculum and we should not wish to see their energies diverted. (However, a comparative element is in fact built into the current Shell Centre/Leicestershire Primary Cal-culators Project.)

   With regard to computers, two issues need highlighting. First, work is in progress on the preparation of software to enable microcomputers to be fully exploited as teaching aids. Secondly, there is the question whether the elements of programming should form a normal part of the secondary mathematics course. Comments on this appear in the chapter.

8. USING RESEARCH

The last chapter is a guide to enable others to find what information they require about research in this field.

   Sometimes one looks to research to find the answer to a specific question, such as whether proof is learnt better through transformations or through Euclidean geometry. But one also expects research activity to be seeking to test instances of more general hypotheses, and so to be increasing our insight into the nature of mathematical learning. We

have aimed in this review to present a reasonably coherent
picture of mathematical learning in accordance with what
research has established up to date. We have tried not to
neglect any major themes. However, we have certainly not
cited all the relevant studies on any one theme. We have,
we hope, given enough references to enable any point to be
pursued more exhaustively, and in these references further
lists of relevant material may be found. Our final chapter,
Using Research, surveys both the range of journals which will
probably be found useful for further study, and also the
general bibliographic references and procedures through which
exhaustive searches of particular topics may be made. It will
be apparent there that while American research material is
generally well abstracted and indexed and hence quite accessible, this is not the case for British research. It may be
useful to consider whether some provision should be made for
this on a regular basis. In this context we are aware of two
further needs which have become clear in the course of this
work: one is for the more regular preparation of survey and
consolidation articles covering areas of research such as
those represented by the chapters of this review; the other
is perhaps for some greater degree of inter-communication
among British researchers in mathematical education as well as
between them and the profession at large. We hope it may be
possible to co-operate with the Committee in formulating
some proposals towards these ends.

9.   CONCLUSIONS AND RECOMMENDATIONS

The main aim of this review is not to formulate recommendations for action but to provide background information for
the Committee. However, it may be helpful to identify those
developments in school mathematics teaching which have
emerged during our study of this material as likely to produce the most significant improvements in effectiveness.

1. *Teaching Methods*

The distinction should be made between skills, concepts and
general strategies as different kinds of teaching objective
and appropriate methods are used for each, viz:

(a) For strategies and general orientation, a substantial
proportion of lessons should consist of solving 'real'
problems and making mathematical investigations;

(b) For conceptual learning, diagnosis of pupils' existing
concepts and the provision of problems based on familiar
situations and concrete materials, with conflicts of
understanding resolved by discussion;

(c) For essential facts and skills, first establishing
understanding and inter-connections with the rest of the
pupil's knowledge. then drill and practice to increase

fluency.

2.   *Curriculum*

The concept of planning courses as if pupils could add
specific techniques to their repertoire at specific times
in the course does not match the results of research. Pupils'
understanding of the basic ideas develops slowly, teaching
needs to be planned so as to offer repeated opportunities
to meet the same basic ideas, variety being provided through
differences of context.

10.   NEEDED RESEARCH

1.   There is a need to continue research of the type reported
in Chapter Six, making 'conceptual maps' of the important
areas of elementary mathematics. This should develop away
from the simple collection of empirical results, as in the
first APU surveys, towards the identification of general
factors determining levels of difficulty - as for example in
Chapter Three, section 3. It might also usefully change its
focus from the mathematics to the learner, attempting to
describe the successive stages of mathematical knowledge and
power through which pupils generally pass.

2.   The research on teaching methods also needs to be continued and developed, particularly towards establishing, if
possible, whole-class or group methods which embody the
attributes of the methods shown in Chapter Seven to be highly
successful with individuals and extending them to a wider
range of common topics.

3.   Also needed is the provision of means by which British
research in mathematical education can develop towards making
a more powerful attack on key problems. This implies greater
co-ordination of the presently fragmented efforts, and
probably the development of a few bigger research groups.

Chapter One
The Mathematics Curriculum

The task of the present Committee of Inquiry is to consider
the teaching of mathematics in relation to the needs of
employment and adult life. It will therefore be concerned
with the suitability of mathematical curricula and with the
effectiveness with which these are taught, and in particular
will no doubt wish to establish priorities for possible
changes or developments in these two aspects. The contribution of this Research Review to the present work will consist
in part of providing material which may help to show what
curriculum choices are feasible in present circumstances - for
example by showing what material is at present not well
learned, and indicating which parts of this material present
particular difficulty - and also suggesting what methods of
teaching seem on the basis of research to be worth encouragement and development. But in order to create a context for
this task we need to make some statements of how we view the
mathematics curriculum today. This chapter, therefore, does
not report the results of specific research but rather seeks
to reflect some of the thinking which has been going on within
the mathematical and educational communities about the nature
of mathematics in its relation to education.

It is not so long since mathematical education in England
consisted, for the lower classes, of training to calculate
accurately with large and complicated numbers, weights,
measures and money, and, for the upper classes, of the rote
learning of Euclid's books. This practice has been subject to
continuous change and now we need to ask again, for this
generation, what kind of knowledge of mathematics is appropriate for a general education. If we took a purely
utilitarian approach to the curriculum, we might suggest that
for pupils of average ability, destined for employment as
clerks, secretaries, nurses, draughtsmen, technicians and the
like, the curriculum would comprise the reading of
instructions, the interpretation of maps and diagrams, writing
for record-keeping, simple tabulation and tallying of quantities, and some practical science or craft. But beyond this
we should certainly wish to offer an educational experience
which would help its products to exercise judgments as
citizens or as parents, and we should also hope to do something to enable a new generation to enter into society's most
highly regarded values and achievements. We should want

11

history, geography, the study of literature, expressive writing, art, music and theoretical science to have a place.

If one accepts the obligation to transmit the skills and culture of civilisation as effectively as possible, the problem of selecting, from the vastness of knowledge, material for the brief years of schooling, is acute. One must seek the most general, the most pervasive, the most distinctive aspects of knowledge. Specifically, one must attempt to establish a structure of ideas which will facilitate the assimilation of further knowledge, and teach the actual skills, strategies and attitudes needed for the acquisition of this knowledge.

In mathematics we should want to reflect as much as possible what kind of subject it is. It is, we suggest, first a means of gaining insights into some aspects of the environment. The form of the growth function of populations, the ways of turning a mattress, the concept of acceleration (and of the decrease in the rate of inflation) and the correct understanding of the statistical 'law of averages' are everyday examples, and any given occupational or leisure situation will furnish many more. Secondly, the general attraction to puzzles and patterns, and the existence of a sprinkling of enthusiastic amateur number theorists, suggest that the capacity for appreciating mathematics as an *art to enjoy* is also present in many people, even if generally suppressed by distasteful school experiences. These two aspects represent the *applied* and the *pure* mathematical approaches. They have been identifiable throughout history as the mainsprings of mathematical activity, and should be recognizable parts of the pupil's experience.

In describing mathematics as 'abstractions, proofs and applications', Alexandrov (1963) claims that it is the high level of abstraction peculiar to mathematics that has given birth to notions of its independence from the material world. However, 'The vitality of mathematics arises from the fact that its concepts and results, for all their abstractions, originate in the actual world, and find widely varied application in the other sciences, in engineering, and in all the practical affairs of daily life.' A comprehensive understanding of the nature of mathematics must recognize this, and mathematical education ought to reflect this relationship with reality.

In spite of its political context, Engels' (1953) identification of 'The numerical relations and spatial forms of the real world' as the object of study in mathematics is a valid starting-point, particularly since he then indicates that these forms and relations may subsequently be studied in their pure form, divorced from their context. This is considered by Krutetskii (1976) as the first basic characteristic of mathematical thought:' ... to isolate form from content, to

12

abstract oneself from concrete numerical relationships and spatial forms.'

Peel (in Servais and Varga 1971) describes mathematics as 'the study of the properties of the operations by which man orders, organises and controls his environment.' Again, it is the properties of the operations which concern the mathematician: they are abstracted from reality, but both the origin and the application of these operations are in the physical environment.

Sullivan (in Steen 1978) and Griffiths (in Wain 1978) also assert the connections of mathematics with the physical world and the intimate relationship between pure and applied mathematics. This insistence on the growth of mathematics from reality, however, does not by any means imply a utilitarian approach to the subject. It certainly does not mean that mathematical activity is to be judged worthwhile only in so far as it has clear practical usefulness. Quadling (in Wain 1978) describes the view that mathematics should be linked to the world of working and living, and should emphasise the transfer of theoretical knowledge to real-life situations; but then points out that many would want to provide some experience of mathematics 'for its own sake' - the recognition of pattern as a source of satisfaction.

This requires both judgments about what is of genuine educational value and considerations of motivation. In a later chapter, research on attitudes will be discussed: some of this suggests that substantial numbers of children in both primary and secondary school believe mathematics to be useful and to be important for employment prospects but this does not make them like the subject, become absorbed by it or even 'hold it in high esteem'. It is not easy to take a long-term view and study mathematics for some envisaged future usefulness: some enjoyment of number and pattern should be available to all.

References

ALEKSANDROV, A.D. *et al.* (1963): *Mathematics: Its Content, Methods and Meaning*. Cambridge, Mass.: MIT Press.

ENGELS, F. (1953): *Anti-Dühring*. Moscow: Gospolitizdat.

KRUTETSKII, V.A. (1976): *The Psychology of Mathematical Abilities in Schoolchildren*. Chicago: University of Chicago Press.

SCHOOL MATHEMATICS STUDY GROUP (SMSG): August 1972, Newsletter No.39.

SERVAIS, W. and VARGA, T. (1971): *Teaching School Mathematics*. Harmondsworth: Penguin Books.

STEEN, L.A. (1978): *Mathematics Today*. New York: Springer Verlag.

WAIN, G.T. (1978): *Mathematical Education*. Wokingham, Berks: Van Nostrand Reinhold.

Chapter Two
Background Psychology

1. INTRODUCTION

This chapter treats two areas of current psychological
research which bear closely on the understanding of mathematical learning. The first concerns memory, and this relates
to our theme in two ways. Problem-solving tasks and most
learning tasks (such as verbal arithmetic problems, or adding
fractions) involve the intake of a number of items of data and
the eventual recognition of some connection among them. This
involves holding the data in the working memory while scanning
it for the required relationships. What the research shows is
that the capacity of this working memory is severely limited
(about three to seven items), and this makes it necessary to
deal with one small subset of the data at a time, then passing
to some other subset - the typical problem-solving search
process. This capacity limitation also leads often to mis-
reading of questions, only part of the information being
extracted. The recognition of this limitation has led to the
development and testing of some specific theories of teaching.
These are treated in the chapter on teaching methods, but the
underlying knowledge about short-term memory is described here.

   The second relevant feature of memory concerns the facts
that long-term memory storage - and subsequent retrieval-
depend essentially on the connectedness of the material stored.
The simple commitment of an isolated item to memory and its

subsequent recall at will almost never occurs. What the
research shows is (a) that what is stored in long term memory
is meaning, not particular forms of words; (b) that for memorization to be effective strategies such as 'rehearsal',
that is, holding the material in consciousness by attending
to it, and 'organizing' or grouping items by recognizing a
connection among them, are necessary; and (c) that if
material to be remembered is intrinsically unconnected, some
mnemonic device is needed, such as fitting it deliberately
into some scheme prepared for the purpose.

In short, this fundamental research on memory supports
and explains the results of the teaching method research
(discussed in a later chapter) in asserting the key importance of meaningfulness and connectedness for successful
learning and retention. The research also shows how deep-
level memory processing depends on the learner's orientation
to understand and consider the material rather than to
memorize it. This clearly has implications regarding the
approach to classroom activity engendered by the kinds of
demand made by the teacher, both immediately and in regard to
the form of testing which the pupils expect to undergo.

The second section of this chapter deals with a series
of experiments by a number of workers which bear on the
effect of context on the difficulty of a logical task.
Mathematical teaching is based on the principle that a given
mathematical concept or relationship - such as 'division' or
proportion - can be learnt in one or more contexts, but
abstracted and transferred to other contexts. For example,
one may learn proportion in the context of mixtures for
concrete or for metallic alloys, and transfer it to dividing
up a pie diagram to show the proportions of time given to
work, leisure and sleep during a day. What this research
shows is that different contexts differ very greatly in the
difficulty they present in the application of the same
mathematical structure. Thoroughly familiar and realistic
contexts are so much easier that it appears that most of the
work needed to apply the mathematical idea has already been
done in that the relationships existing in the material
already exist in the mind and have only to be chosen. For
the classroom this implies both that the use of a *genuinely*
familiar context may give a first access to the idea for many
more pupils, but also that the abstraction of the idea in a
sufficiently detached form to be applicable to unfamiliar
contexts may present difficulties of a quite different order.

The main research was actually performed with a particular logical task, the Four-Card Problem, but it may stand as
a fairly typical example of a mathematical task. Before
discussing this, we consider other work which shows the high
level of difficulty which logical tasks may possess. The
kinds of deduction which are crucial to mathematical proof -
the distinction between one-way implication and two-way

equivalence, for example - are much more difficult than has
generally been assumed. If they are to be successfully
learned by even the more able secondary pupils, considerable
teaching effort will be needed. But it is also apparent that
they are called into play less frequently than one might
imagine in most mathematical activity.

2. MEMORY

Introduction

The models that have been developed over the last 20 years or
so to describe memory have generally classified memory into
three levels, each with their own storage systems: sensory
memory, short-term memory and long-term memory. Sensory
memory does not require the paying of attention; stimuli
(such as the background sound of a radio, or the voice of a
teacher) can be stored directly, in their raw state, without
processing. However, material in the sensory memory stores
fades in a matter of one or two seconds, and is therefore
quickly lost, unless the subject attends to it within that
period of time, thereby transferring the material to short-
term memory. Here, the memory trace may last for 20 or 30
seconds and can be maintained for longer by continued
attention: for example, an unfamiliar telephone number read
from a directory can be remembered long enough to be dialled
by repeating the string of digits one or more times. Such
repetition, or rehearsal, can also help to establish material
in long-term memory, although this will depend on the nature
of the material and the features that are noted. In short-
term memory the features stored tend to be superficial - for
example, the sound or visual appearance of a word - whereas
long-term memory is more concerned with meaning.

Once an item is stored in long-term memory it tends to
be forgotten very slowly or not at all (though this does not
imply that it is necessarily easy to retrieve); in contrast,
an item will fade from short-term memory within 30 seconds
or less, unless it is kept in consciousness. Moreover the
capacity of short-term memory is severely limited, so that
existing items are also easily forgotten by being displaced
by oncoming items.

Craik and Lockhart (1972), in a review of the evidence
for this 'multistore' approach to memory, provide a useful
summary (shown below) of the commonly accepted features of
the three kinds of memory store, in cases where the material
to be remembered is verbal.

Commonly Accepted Differences Between the Three Stages of
Verbal Memory (See Text for Sources)

Feature	Sensory Registers	Short-term store	Long-term store
Entry of Information	Preattentive	Requires attention	Rehearsal
Maintenance of Information	Not possible	Continued attention Rehearsal	Repetition Organization
Format of information	Literal copy of input	Phonemic Probably visual Possibly semantic	Largely semantic Some auditory and visual
Capacity	Large	Small	No known limit
Information loss	Decay	Displacement Possibly decay	Possibly no loss Loss of accessibility or discriminability by interference
Trace duration	1/4 - 2 Seconds	Up to 30 seconds	Minutes to years
Retrieval	Readout	Probably automatic Items in consciousness Temporal/Phonemic cues	Retrieval cues Possibly search process

At the same time, Craik and Lockhart argue that sufficient
evidence has now accumulated to show that this view of memory
as being composed of distinct stores with clearly differen-
tiated properties is no longer adequate, and they suggest that
it might be more profitable to abandon the multistore approach,
despite its intuitive appeal, and to focus on how different
ways of attending to material affect the degree to which the
material is memorized (see also Craik and Tulving, 1975,
below). However, for the present it is proposed to stay with
the multistore approach since it provides a useful framework,
albeit an oversimplified one, for discussing different aspects
of memory and, as such, has gained general acceptance among
psychologists.

Short-term memory

Though only a small part of mathematics teaching may be concerned with rote-learning, it is clear that the effects of at

least some aspects of what children experience during a
mathematics lesson are intended to be more or less permanent.
If therefore it is the case that material held in short-term
memory fades in a matter of seconds, unless deliberately
held in consciousness, it might seem at first sight that
research into short-term memory *per se* (as opposed to the
transfer of information from short to long-term memory stores)
is of little relevance to the problems of teaching. The
stark and remote nature of the tasks used to investigate
short-term memory adds weight to this view: thus, typically,
tasks such as digit-span tests are used, in which several
digits are presented in rapid succession orally or by brief
exposure on a screen, and the subject is tested to see how
many he can recall, immediately and in the correct order.
But in fact short-term memory plays an important role in
all tasks where several attributes or items of information
have to be considered simultaneously, such as mental arithmetic, problem solving, the understanding of complex concepts,
constructing or following an explanation or argument - that
is, in most learning tasks.

If digit-span is taken as a measure of short-term memory
capacity, then the use of immediate-recall tests with children and with adults has shown it to have two important and
well-attested properties. One is that short-term memory
can carry only a very small number of items - given as about
seven for adults, in a seminal paper by Miller (1956); the
other is that the number of items increases with age. A
rather literal interpretation of these findings is to assume
that short-term memory storage consists of a finite number
of 'boxes' or 'slots', each of which can store one item of
information, and that the number of slots increases as the
child matures. However, it has also been shown that span
depends on the familiarity of the items (e.g., Crannell and
Parrish, 1957), with strings of words being generally more
difficult to recall than letters, and these in turn being
more difficult than digits, which suggests that the 'number
of items' view is too simple. One way of accommodating this
variation is to assume that items are stored in 'chunks',
where the number of items in a chunk is dependent on the
meaningfulness, or relatedness, of the items. Some form of
chunking undoubtedly occurs in certain recall tasks: for
example, Craik and Masani (1969) have shown that young children can reproduce strings of as many as 20 words in a span
test when the words form a sentence. However, it is not
clear why such chunking should occur more readily for random
digits than for letters, say; also, invoking chunks to
support the notion of discrete memory slots is essentially
circular, unless some way can be found of defining them
independently of the results of span tests; finally, to
assume that complex meanings as well as simple physical features (such as the sound or physical shape of a letter or
digit) can be stored in short-term memory lessens and thereby casts doubt on, the distinction between short- and long-

term memory.

Another explanation for the increase of span with age
is to assume that, though the number of memory slots (if
they are deemed to exist) remains constant, older children
have developed better strategies for remembering material.
One such strategy might be rehearsal, whereby material is
maintained in consciousness or transferred to long-term
storage by continued repetition; another might be to
organize the material into smaller groups (as is commonly
done with telephone numbers for example), not necessarily as
a way of chunking but as a way of emphasizing the position
of items in a string. There is strong evidence to suggest
that children's *general* use of such strategies (and also their
awareness: see e.g., Kreutzer *et al.*, 1975) does increase
with age. For example, Flavell *et al.*, (1966), in a series
of *delayed* recall tests with children between the ages of
five and ten years, found that the older children were
better able to recognize which subset of pictures, out of a
total set of seven pictures, the experimenter had previously
pointed to. Since the delay was typically of about 15
seconds duration, and since the total array of pictures was
hidden during this period, it is reasonable to assume that
children who were able to use a rehearsal strategy would
have been at a considerable advantage. It was also found
that very few of the five-year-olds, but most of the ten-
year-olds, showed signs of verbal activity (e.g., lip movements) during the delay period, and that those who did
show such activity performed better on the test. In an
extension of this investigation, Keeney *et al.*, (1967) found
that when children who appeared to show no signs of rehearsal were instructed to whisper the names of the pictures
during the delay period, their retention accuracy matched
that of the children who rehearsed spontaneously. But
interestingly, these children abandoned the rehearsal
strategy, despite its proven benefits, when the instruction
to whisper the names was no longer given.

In span tests, where the required recall is immediate,
there is less opportunity for rehearsal than in the delayed-
recall tasks described above. However, there is evidence
to suggest that a *certain* amount of rehearsal does occur:
commonly, subjects remember the first and last few items in
a list better than items in the middle (the so-called
'primacy' and 'recency' effects). As far as the last items
are concerned (recency), it can be argued that these are
retained more successfully *not* because of rehearsal, but
simply because they have had less time to fade from short-
term memory. Support for this view comes from Craik (1970)
who found that when subjects were given a secondary task
that delayed immediate recall, the last items were recalled
worst of all (negative recency). On the other hand,
rehearsal and, in particular, *cumulative* rehearsal (*a, ab,
abc,* etc.), which gets progressively more difficult as the

list of items gets longer, provides a possible explanation
for the better recall of the first few items (the primacy
effect). Direct support for this comes from Kingsley and
Hagen (1969) who found that when five-year-olds (who can be
assumed not to rehearse spontaneously) were trained to use
cumulative rehearsal, recall of the first few items was
markedly improved. Hagen and Kingsley (1968) presented
children aged four to ten years with a series of pictures,
and asked the children to say aloud the name of each picture
as it was presented. This can be assumed to *interfere* with
cumulative rehearsal (in which not just the current picture
is named but all the previous ones are named again), and this
is confirmed by the fact that explicit naming reduced the
primacy effect for the ten-year-olds but not for the younger
children, where the primacy effect stayed at about the same
level. In contrast, naming enhanced recency for all age
groups.

Given then, that short-term memory span increases with
age, that the use of rehearsal strategies increases with age,
and that the primacy effect in span tests seems to be due, at
least in part, to rehearsal, it is reasonable to assume that
there is a link between the development of span and of
rehearsal. However, a study by Huttenlocher and Burke (1976)
suggests that other factors must also be operating. They
examined the primacy effect for childred aged four, seven,
nine and 11 years on an immediate-recall test of ordered
digits presented orally. Though span increased with age, and
though the primacy effect (the better recall of the first few
digits) was apparent for each age group, they were unable to
discern any marked *increase* in the primacy effect for the
older age groups. This suggests that rehearsal cannot be the
sole cause of primacy, and more importantly, cannot be the
sole cause of the increase in span. Instead, Huttenlocher
and Burke propose that span is dependent on the degree of
familiarity with the items being encoded and on ordering
ability, both of which increase with age. With respect to
familiarity, Huttenlocher and Burke cite a number of studies
that have shown, for example, that span is greater for digits
(of which there are only ten) than for English letters (of
which there are 26); for common than for uncommon or nonsense
or foreign words; and also increases with decrease in background noise. As for ordering ability, there is substantial
evidence to show that this improves with age (e.g., Piaget
and Inhelder, 1969); moreover, it is likely that children more
easily remember the order of items near the beginning and end
of a list because their position is more distinct. The
importance of positional cues is further demonstrated by the
fact that when Huttenlocher and Burke slightly increased the
time interval between the presentation of two of the middle
items in their tests, thereby effectively increasing the
number of 'end' points, recall of the middle items improved.

Rather than regarding short-term memory as consisting of

a fixed number of slots for storing information (these slots increasing with age), Huttenlocher and Burke invoke the notion of 'attention' or 'processing capacity'. With respect to item-familiarity, they then argue that this fixed amount of attention is used both to *identify* incoming items and to hold them in active memory; this means that if the items are more difficult to identify (because unfamiliar, for example), less attention is available for maintaining items in short-term storage (or, as will be discussed in the next section, for committing them to long-term memory). Support for this fixed-attention hypothesis comes from Rabbitt (1968) who showed that items in the first half of a list became more difficult to recall when items in the second half were presented in noise.

The concept of fixed 'attention' provides a more comprehensive description of short-term memory rather than focusing on storage capacity alone. However, it is perhaps even more useful to talk about a limited *processing* capacity when interpreting the results of span tests, since it can be argued that these tests are not just a matter of identifying items and 'attending' to them: other mental processes are involved (such as the selection, retrieval and use of information stored in long-term memory) which may affect the *way* in which the items are identified (encoded) and attended to, thereby determining not only the length of time that the items are circulated or maintained in storage, but whether they are held in short- or long-term storage. This suggests also that it might be better to use the term 'working memory' rather than 'short-term memory', and some writers prefer this term (see, for example, Baddeley and Hitch, 1974).

In summary, research using immediate recall tests (span tests) has shown that short-term memory capacity (whether this is called processing or storage capacity) is severely limited. On the other hand, for a given kind of item (digits, letters, etc.), span increases with age. One explanation for this assumes that short-term storage consists of discrete slots which increase in number with age; this seems to be an oversimplification which ignores the complex processes involved in encoding, maintaining, storing and retrieving information. The awareness and spontaneous use of strategies such as rehearsal increase with age and some use can be made of these even when the recall required is 'immediate'. More important in the context of immediate recall is the child's ordering ability, which is known to increase with age, and the degree of familiarity with the items being encoded, which is likely to influence both the amount of processing capacity devoted to identifying the items and the way in which they are processed.

Processing capacity in relation to arithmetic

Hitch (1978) has examined how what he calls 'short-term working memory' relates to mental arithmetic. In arithmetic items, it seems likely that the initial information (the digits to be added, say,), unless written down, is held in working (or short-term) rather than long-term storage. This in turn suggests that errors are likely to increase as the amount of written information decreases. To test this, Hitch gave subjects addition items where the written information varied as in the four conditions below. In conditions T, B and O, the missing digits were given orally and writing them down was not allowed, though subjects were asked to write each partial result as soon as it was computed.

Condition TB	Condition T	Condition B	Condition O
352	352		
+259	+___	+259	+___

The error-rate for the four conditions were .03, .08, .07 and .22, which is in the direction predicted.

It also seems likely that the *longer* material is stored in working memory the more likely it is to be forgotten. To test this Hitch used mental additions such as 324 + 253, where there is no carrying, so that each of the steps involved (adding the units, the tens, and the hundred) are of the same order of difficulty, but where the left-hand digits have to be held longer (on the assumption that subjects use a units-tens-hundreds strategy). The relative size of the observed error-rates was as expected: 9 per cent for the units figure: 18 per cent for the tens; 31 per cent for the hundreds.

Hitch argues that the errors observed in these and other experiments that formed part of the study can be adequately explained by the loss of information in working memory through a gradual 'decay' rather than through an 'overflow' of storage capacity, as proposed, for example, by Lindsay and Norman (1972) and Hunter (1964). However, he accepts that decay rate may well be a function of instantaneous load on working storage and that the effect of storage overflow might be seen more directly in problems where greater storage loads are involved.

Long-term memory

There is substantial evidence to show that information is better remembered if it is meaningful. This is not to say that non-meaningful material, or material that is not *processed* meaningfully, cannot be stored in long-term memory: however, it is likely to be less well retained; perhaps more importantly, non-meaningful material is more difficult to *retrieve* from long-term storage because fewer associations have been formed to act as retrieval cues.

Craik and Lockhart (1972) review a number of studies in which subjects were oriented to process material in different ways, and they argue that it is 'depth' and 'elaborateness' of the processing that determines how well material is subsequently recalled. In one such study (Tresselt and Mayzner, 1960), subjects were presented with a series of words and asked to (a) cross out vowels or (b) copy the words or (c) judge whether the word related to the concept 'economic'. On a subsequent free-recall test, four times as many words were remembered under condition (c) and twice as many under condition (b) than under condition (a). In another study (Shulman, 1971), subjects had to scan a list of words for features that were either structural (e.g., words containing the letter A) or semantic (e.g., words denoting living things). On a subsequent unexpected recognition test, performance in the meaningful (semantic) condition was significantly better than in the structural condition, even though scanning time was about the same for both. A similar result was found by Hartley (1980) in a study involving students' recall of statistical material.

Craik and Tulving (1975) conducted a series of ten experiments in which the relationship between depth of processing and recall/recognition was investigated. In one of these experiments 24 paid college students were presented individually with a series of 60 words. For each word they had to answer one of three kinds of yes/no questions representing three levels of encoding:

sturctural (e.g., Is the word in capital letters?)
phonemic (e.g., Does the word rhyme with WEIGHT?)
semantic (e.g., Would the word fit 'He met a .... in the street?')

The time taken to answer each question (the 'response latency') was measured, and after the 60 trials were completed, subjects were given an unexpected recognition test consisting of a list of 180 words that contained the original 60 words.

As the adjacent figure shows, recognition increased substantially for deeper level of processing. However, this effect was far less strong for 'no' questions than 'yes' questions. Also, it was found that there was an almost perfect linear relationship between the time spent on a question and recognition rate. which suggested that memory performance might simply be a function of processing *time* rather than *level* of processing (but see below).

It can be argued that the 'yes'

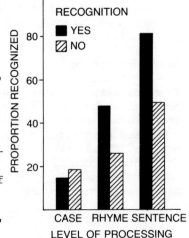

24

questions produced better results than the 'no' questions because in the former case the given word could be more fully integrated with the question, thus producing a richer, more elaborate encoding. This was confirmed in a subsequent experiment, using questions like 'Is the object bigger than a chair?', where words producing a 'yes' (e.g., HOUSE) or 'no' (e.g., MOUSE) response can be equally easily integrated and where no difference in recall was found between words in the 'yes' and 'no' conditions.

To show that recognition was not simply a matter of processing time an experiment was undertaken using the same semantic material as before (e.g., Would the word fit 'The man threw the ball to the ....'?) but using a more difficult non-semantic task what would take longer to perform. This task consisted of showing a pattern of V's and C's such as CCVVC (where V = vowel and C = consonant) and then asking whether a subsequently presented word (e.g., BRAIN) fitted this pattern. As expected, the response latencies were far grater on this task than on the semantic task, but on the memory test the semantically processed words were again better recognized:

Response type	Level of processing	
	Structure	Sentence
Response Latency (sec)		
Yes	1.70	.83
No	1.74	.88
Proportion recognized		
Yes	.57	.82
No	.50	.69

In some of the experiments free-recall rather than recognition tests were used, but the pattern of results was the same, with the semantically encoded words remembered best. When subjects were *forewarned* that they were to be given a memory test recall improved at all levels of encoding (structural, phonemic, semantic), which suggests that when subjects are instructed to 'learn' a list of items they tend to initiate their own encoding operations. However, Craik and Tulving make the important point that these encoding operations may not be the most effective, since words in the semantic condition were still remembered best.

It is widely accepted that a way of improving recall, both in the short- and the long-term, is by rehearsal: rehearsal is a way of maintaining an item in short-term memory, and the longer an item is maintained the more likely is it to enter long-term storage. However, studies by for example, Craik and Watkins (1973) and Woodward *et al.* (1973) suggest that the crucial factor determining long-term storage is how deeply or elaborately an item is processed during rehearsal rather than the length of time *per se*, that the item is maintained.

In one of the experiments devised by Craik and Watkins, subjects were told to listen to a series of words, and to report the last word beginning with a certain letter. Subsequently, subjects were unexpectedly asked to recall as many words as possible, whether or not they began with the specified letter. If the series consisted, say, of 'Daughter, oil, rifle, garden, grain, table football, anchor, giraffe ...', and the critical letter was 'g', it can be argued that subjects would ignore the first three words, hold 'garden', replace it by 'grain' and replace this by 'giraffe' which is the word that would be reported if the remaining words were non-critical. The greater the number of words, N, between one critical word and the next, the longer the first of these critical words is maintained in short-term memory. Therefore, if length of time that an item is maintained is crucial for transfer to long-term memory, more words with high than with low N values should be remembered on delayed recall. However, as can be seen from the 'rows' in the table below, N value had no obvious effect (N non-significant, $p > .10$). Craik and Watkins conclude that 'time in short-term store will only predict later long-term store performance when the subject has used the time to code the item elaboratively' (p.603).

PERCENTAGE RECALL AS A FUNCTION OF EXPERIMENTAL CONDITION N VALUE, AND PRESENTATION RATE

| Condition | Presentation Rate | N Value | | | | | | | | | |
		0	1	2	3	4	5	6	8	12	Mean
Replaced	Slow	12	13	22	10	21	19	19	18	19	17
Critical	Medium	10	15	22	12	14	19	09	12	11	14
Word	Fast	14	07	11	06	06	14	09	16	15	11
	Mean	12	12	19	10	14	17	13	15	15	14
Reported	Slow	19	20	20	20	31	39	22	26	28	25
Critical	Medium	20	22	19	19	31	26	20	28	20	23
Word	Fast	26	15	22	26	20	31	19	11	20	21
	Mean	22	19	20	22	28	32	20	22	23	23

A study by Woodward *et al.* (1973) confirmed Craik and Watkins' (1973) findings with respect to delayed *recall*; however they did find that rehearsal time enhanced subsequent *recognition* of items. Woodward *et al.* presented lists of words with a delay of 0, 4 or 12 seconds between each word followed by a cue (a red or green light) telling subjects to remember (R) or forget (F) the current word. Subjects were then given an immediate (free) recall test, a delayed recall test and finally a delayed recognition test, the results of which are shown below. Woodward *et al.* argue that during the delay period, since subjects did not know whether they would be asked to remember the word or not, subjects would rehearse the word in a rote, non-associative manner rather than in a more constructive fashion. This is confirmed by the first two graphs below, which show that rehearsal time had no effect on recall ($p > .25$ and $p > .05$ respectively). However, the third graph suggests that rote learning does have *some* effect on long-term memory, since *recognition* of items increased with rehearsal time ($p < .05$).

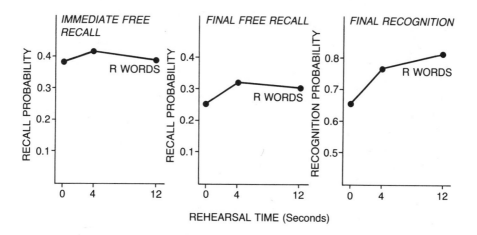

REHEARSAL TIME (Seconds)

Aids to recall

The previous section has shown that information is more likely to be stored in long-term memory, and more easily retrieved, if the material is 'integrated' or 'associated' with other material, i.e. considered in a meaningful rather than in an isolated, rote manner. There are a number of memory aids based to a greater or lesser degree on this principle.

A procedure where this principle is minimally applied is 'first letter cueing' where the letters of the alphabet are considered in turn as possible first letters of a word that the subject is trying to recall. A study by Gruneberg and Monks (1974) suggests that this procedure can be quite effective for retrieving material that the subject is confident

27

he knows but has temporarily forgotten. However, it is an open question whether it is the specific nature of the procedure that makes it effective or simply the time spent on recall.

A related procedure is to use first-letter mnemonics as an aid to the learning and subsequent recall of lists such as 'Richard Of York Gave Battle In Vain' for the colours of the rainbow. However, according to Gruneberg (1978), the results of studies that have investigated the effectiveness of such mnemonics have tended to be inconclusive or negative. On the other hand, even if mnemonics *per se* are not particularly effective, the time spent devising and using them may well be beneficial; also, clinical evidence cited by Gruneberg suggests that in stressful situations, such as examinations where efficient recall is vital, the use of mnemonics may greatly reduce anxiety.

Another procedure for remembering lists, which involves far more elaborate associations than using first letters, is the method of 'loci'. Here the items to be learnt are related one-to-one to a known list of cues consisting of memory images of geographical locations, such as landmarks on a familiar journey, or parts of a room, or even parts of the body. The key to the method is to set up an *interaction* between the items and the cues that is as vivid as possible. Research suggests that the method can be highly effective: for example, Groninger (1971) found that a group who had been taught to use the method could recall 20 out of 25 words in their correct positions in an ordered list after a five-week period, compared to ten out of 25 words for a group who had no instruction.

The 'hook' or 'peg' system is an extension of the loci method. Here a poem, such as 'One is for Bun, Two is for Shoe, Three is for Tree, ...' etc., is first learnt, and then the words that are to be memorized are associated as vividly as possible to Bun, Shoe, Tree, etc., respectively. Several studies (e.g., Bugelski, 1968) have shown this method to be far superior to ordinary rote learning.

The 'successive comparison technique' does without hooks. Instead the words to be learnt are associated with each other, by constructing images that relate the first and second word, the second and third, the third and fourth, etc. In a variation of this technique Bower and Clark (1969) instructed subjects to learn lists of ten words by making up a story linking the words in order. Subsequently, 93 per cent of words were recalled in their correct order, compared to 14 per cent for a control group.

The development of processing capacity and the Piagetian stages

A number of attempts have been made to relate Piaget's stage theory of cognitive development to the development of short-term memory capacity, by describing tasks at the different stages in terms of the number of elements (or concepts or schemes) that have to be considered simultaneously.

McLaughlin (1963) proposed that the number of concepts that need to be coordinated in preoperational, concrete and formal operational tasks is 2, 4 and 8 (2^1, 2^2, 2^3) respectively, and he suggests that a child would need a digit-span equal to the number of concepts to cope with tasks at a given stage. Halford (1978) also argues that a span of two is required for preoperational tasks, and a span of four for concrete operational tasks, but he suggests that formal tasks require a span of six rather than eight. In the theory developed by Pascual-Leone and his co-workers (see, for example, Case, 1974) the emphasis is more on the child's processing capacity than storage capacity. Thus rather than using a straightforward span test, the child is tested on immediate recall of information that first has to be operated on in some way. For example, the child might be asked to recall the number of elements in several sets which he has had to count; according to Case (1978), children below the age of five years can only remember the number of elements in one set, whereas at the age of seven or eight, say, they can remember the elements in as many as three sets. The table below (from Case, 1974) shows the presumed relationship, according to Pascual-Leone's theory, between age, Piagetian developmental level and processing capacity (or 'M-power'). The constant e refers to the mental effort (or energy, capacity or space) required to attend to the specific instructions of a task (when these are easily understood and remembered), whilst the numeral represents the maximum number of schemes that have to be considered simultaneously to solve a task at a given Piagetian sub-stage and whose value, for an individual child, can be determined by the kind of test referred to above.

Age	Developmental substage	M-power
3-4	Early preoperations	$e+1$
5-6	Late preoperations	$e+2$
7-8	Early concrete operations	$e+3$
9-10	Middle concrete operations	$e+4$
11-12	Late concrete-early formal operations	$e+5$
13-14	Middle formal operations	$e+6$
15-16	Late formal operations	$e+7$

There is a certain amount of empirical evidence to support Halford's and especially Pascual-Leone's theory (e.g.,

Halford, 1978, and Case, 1974, page 548, footnote 4). However, at the present time at least, their practical value for determining the difficulty of a task is limited, for two reasons; one, the analysis required to determine the schemes relevant to a particular task can be extremely complicated (e.g., Pascual-Leone and Smith, 1969) and ambiguous (e.g., Lawson, 1976); two, factors other than just the *number* of schemes that have to be coordinated can affect task diffi- culty: for example, the familiarity of a cue for a given individual, and the salience of a cue for a given type of task (e.g., how clearly the cue is presented) can both affect the amount of mental effort required to take note, make sense and make appropriate use of the cue; also studies of what Witkin has called 'field dependence-independence' (Witkin *et al.* 1962) suggest that individuals with the same processing capacity may differ in the extent to which they make use of this capacity and in their ability to overcome the attraction of misleading cues (Case, 1974).

Despite these serious practical limitations, these 'neo- Piagetian' theories do have important implications for teach- ing; at the very least, they underline the fact that the complexity of a task, in the sense of the number of schemes that need to be considered at any one time, can have a crucial effect on the task's difficulty. More fundamentally, the theories suggest that the child's processing capacity imposes a limit on the kind of task that the child can be taught to solve or understand (though it might be possible to simplify a task in some way, so as to reduce its processing demand). This means, in contrast to a learning theory such as Gagne's, that to teach a complex concept or skill it is necessary to ensure not only that the child has acquired all the pre- requisite concepts or skills but that the number of these that have to be integrated at any one time is kept to a certain minimum, depending on the level of development of the child's processing capacity.

Case (1975) gives several examples of children who can cope successfully with the separate components of a task but are unable to coordinate these components. In the task illustrated below, it is necessary both to avoid blind alleys and to judge which of several routes is the shortest.

'Terry (5 years, 0 months) goes the shortest way consist- ently in simple maze situations; however, on transfer problem 6 (below) he avoids the blind alley and completes the trip by a longer route than is necessary. I show him three possible ways to get to the goal and ask him to pick the shortest. He picks the correct one' (Case, 1975, p.71).

Thus Terry knows what it means to pick the shortest path, and also how to avoid dead ends, but he cannot hold both require- ments in his mind while searching for his route and satisfy them both.

30

Transfer problem for maze learning. The instructions are as follows: "These are islands. These are bridges. This is water. What is the shortest way from here (X) to the house without going into the water?"

Collis provides similar examples in a mathematical context. In one study, Collis (1975a) identified children who were able to cope successfully with simple items like 2 + 3, but whose working broke down when an extra element and operation were introduced, as in 2 + 3 + 4, say. Typically, these children responded in the following kind of way:

E : What number does 2 + 3 + 4 equal?
S : 2 + 3 = 5 and (pause) what was the other number?
E : 4
S : Ah yes! Now, 2 plus (pause) what is the sum again?

In another study, Collis (1975b) contrasts children who try to solve an item like 'find y, if $y = b$ and $y + 2b + = 90$' by giving numerical values to b and who eventually get lost in a

31

series of trials', with children at a higher level of cognitive development, who argue in the following, far more
efficient manner, '*b* is a number; 2*b* is twice that number and
thus twice *y*; *y* and 2*y* make 3*y* and thus *y* = 30' (ibid, p.46).
There is an interesting paradox in this example, in that the
second group of children, who can reasonably be assumed to
have a higher processing capacity, are able to choose a strategy which seems to be less complex (though more sophisticated)
than the trial-and-error methods of the first group of
children. It might be argued that the second group's success
is simply due to a greater familiarity with letters representing unknown numbers. However, this begs the question of how
such familiarity comes about: in part it may be due to
greater experience with letters used in this way, but it can
also be argued that the ability to use letters as numerical
entities in their own right, or, rather, *learning* to use
letters in this way, requires a greater processing capacity
than working with numbers *per se*. (For example, it might be
the case that initially children support their work with
letters by referring to known numbers and the way they
behave.) This raises the interesting possibility, in
generalized arithmetic and other areas of mathematics, that
an increase in processing capacity may sometimes have a twofold beneficial effect: not only may it give children access
to more effective strategies, but these strategies, once
understood, may in some cases require less processing
capacity than the more primitive strategies they displace.

3. THE EFFECT OF CONTEXT ON LOGICAL THINKING

Introduction

Much of the research into logical thinking has been concerned
with logical implication, and only such research will be
considered here.   In particular, two kinds of tasks involving
logical implication will be discussed: one is concerned with
making inferences: for example, given the implication 'If
the car is shiny, it is fast' and the premise 'the car is
fast', subjects can be asked whether the conclusion 'the car
is shiny' is true, untrue, or whether there is not sufficient
information to make a valid inference. The other kind of task
has become known as Wason's selection task, or the four card
problem: in this, four cards are shown, bearing on their
visible side a consonant, a vowel, an odd number and an even
number respectively; the statement 'Every card with a vowel on
one side has an even number on the other side' is given; the
question is which cards must be turned over to determine
whether this statement is true. Thus the subject has not to
make a direct inference but to consider what evidence is
necessary and sufficient for settling the given statement -
a harder task.

Making inferences

The first kind of task has been investigated by O'Brien in a series of studies with children, adolescents and college studens (O'Brien *et al.*, 1971; O'Brien, 1972, and 1973 respectively). O'Brien used written items in which the logical form of the inference was varied, as in the items below:

1. MODUS PONENS
    If the car is shiny, it
    is fast
    The car is shiny.
    Is the car fast?
    (a) Yes
    (b) No
    (c) Not enough clues

2. CONTRAPOSITIVE
    If the car is shiny, it
    is fast
    The car is not fast
    Is the car shiny?
    (a) Yes
    (b) No
    (c) Not enough clues

3. INVERSE
    If the car is shiny, it
    is fast.
    The car is not shiny.
    Is the car fast?
    (a) Yes
    (b) No
    (c) Not enough clues

4. CONVERSE
    If the car is shiny, it
    is fast.
    The car is fast.
    Is the car shiny?
    (a) Yes
    (b) No
    (c) Not enough clues

In the sense that the first two forms allow definite con- clusions, they can be described as 'closed', and in all his stud- ies O'Brien found that these were substantially easier than the 'open' forms three and four, in which there are 'not enough clues' to make a definite infer- ence. This can be seen clearly from the adjacent graph (O'Brien, 1972), in which the subjects were girls aged 14-17 years with a mean 1Q of about 110.

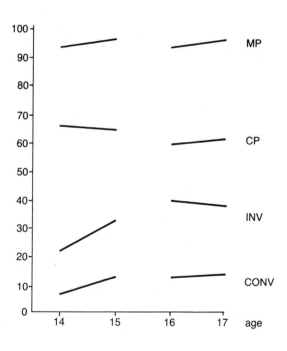

It is noticeable also from the graph that there is little improvement in performance with age, and O'Brien (1973) found that the forms were of comparable difficulty even for college students who were taking a course in logic (96 per cent, 64 per cent, 44 per cent and 18 per cent correct for forms one to four on a similar version of the test).

These results suggest that many subjects are using what O'Brien calls 'child's logic', rather than 'maths logic', by interpreting the implication as an equivalence: i.e., if the first premise is true the second is true, and vice versa; and similarly if the first is false the second is false, and vice versa. Knifong (1974), using Piaget's terminology, calls this transductive logic.

The over-all pattern, that the closed forms are easier than the open ones, is not surprising; it is hard, particularly for younger children, to accept 'not enough information' as a conclusion. But it is an interesting feature of O'Brien's results that there are also substantial facility differences *within* the closed and open forms. On closer examination of the data, O'Brien (1973) found that a sizeable minority consistently gave the 'not enough clues' response to both the contrapositive and inverse forms (two and three) but without using this response indiscriminately in that they still gave definite answers to form one (correct) and form four (incorrect). In effect, these children were giving definite answers when presented with either one of the original premises, but were unwilling to make any inference from negative instances ('the car is *not* shiny' or 'the car is *not* fast'). This way of interpreting implication seems less primitive than the 'child's logic' referred to above, although it is not possible to be sure of this from O'Brien's analysis.

O'Brien also investigated the effects of different contexts on logical reasoning ability, by using the four context-types shown below (O'Brien, 1972).

Basic sentences

Test I - Causal
1. If Joe will play, the Tigers will win.
2. If the horse does jump, Jim will get hurt.
3. If Sue screams, Jim will Jump.

Test II - Class inclusion
1. If the car is shiny, it is fast.
2. If that dog is smart, it is brown
3. If the horse is red, it is big.

Test III - Nonsense
1. If crayons are fobel, cats are raffish.
2. If stars are fantic, planets are monive.
3. If hippos are fruckly, tigers are gandy.

Test IV - Random
1. If the hotel is pleasant, Jane is young.
2. If the fire truck is old, the car is blue.
3. If the square is red, the toy is big.

The effects of context on item difficulty differed from one logical form to another, which makes the effects difficult to interpret. However, over all, the effects were small, which at first sight is rather surprising: it might have been expected that the nonsense and random contexts would have been more difficult because they appear to be less meaningful. However, it can be argued that the causal and class inclusion items, despite their appearance, are not very meaningful either: though in these cases the premises *could* be connected (a smart dog *could* be brown), the connections are not necessarily plausible. In contrast, it is interesting to consider the implication 'If it is Jim's car, it is red' which comes from another study of O'Brien's (O'Brien *et al.*, 1971). Though O'Brien, again, simply regards this as a class inclusion item, the relationship here is far more plausible: it is possible to imagine a person called Jim whose car is red. Moreover, it is possible to evoke or imagine a familiar situation, such as walking down the street, in which there is a red parked car that is Jim's, there are other red cars that are not Jim's, there are not-red (blue, say) cars that are not Jim's, but there are *not* not-red cars that are Jim's. Given this familiar 'frame', it should be a relatively straight-forward matter to check off whether, given one of the premises, the other *will* be true, will *not* be true, or *may* be true. Using the symbols J and R for the premises 'Jim's car' and 'red car', and J and R for their negation, the argument being put forward here is that the likelihood of a subject making correct inferences, given the implication J R, will depend on how easily the combinations J.R, J.R, J.R, and J.R are evoked, and how readily it is recognized that, of these, J.R is false.

A study by Hadar (1978) suggests that extensive training with concrete materials (in which children were provided with ample opportunity to recognize that there are not always enough clues to make a valid inference - but without being given a rule) can produce a substantial improvement in performance on a written test similar to that used by O'Brien. With 104 11-year-olds, she found that 81 children (78 per cent) achieved a net gain in the correct use of the 'not enough clues' response, compared to their performance on a pre-test. However, most of these children (65) also used this response more often after training than before in cases where it was inappropriate (e.g., the contrapositive logical form) so that it is unclear whether their net gain may at least in part have been due to overlearning the 'not enough clues' response, or whether even this inappropriate use may have represented a genuine increase in insight. Also, it is not clear how long the improvement in performance would have persisted, as no delayed post-test was given.

Wason's selection task

Turning now to Wason's selection task, in its original form this task was as follows (Wason, 1966): subjects are presented with four cards, showing, respectively, A, D, 4, 7; from previous experience it is known that every card, of which these are a subset, has a letter on one side and a number on the other. Subjects are then given the rule 'If a card has a vowel on one side, then it has an even number on the other side', and are told 'Your task is to say which of the cards you need to turn over in order to find out whether the rule is true or false'.

To solve this task, it is necessary to turn over A and 7 in order to check that the combination vowel x odd does not occur. The other two cards do not matter: from the rule, D (non-vowel) can have an even *or* an odd number on the other side, the 4 can have a vowel *or* a non-vowel. Thus the key to the task is to consider only those cases which might falsify the rule. According to Popper (1959) this notion of falsification is crucial to an understanding of the status of scientific theories: a theory can never be proved, only disproved. However, this does not mean that normal scientific activity is exclusively, or even primarily, concerned with falsification: indeed, quite the opposite. To illustrate this point, it is of interest to consider one of Piaget's formal reasoning tasks, such as 'Flexible Rods'. Piaget himself evokes a logical model to explain children's success on this and similar tasks to be found in *The Growth of Logical Thinking* (Inhelder and Piaget, 1958), and it is perhaps this model, more than anything else, that has prompted so much research into logical thinking. The model suggests that children at the formal operational stage have an understanding of formal logic, and, by implication, an understanding of the 'falsification principle' required by Wason's task. However, Wason's studies show that this clearly is not so: but nor does it follow from the behavioural descriptions provided by Piaget himself. In the flexible-rods task, children are provided with a number of rods, which vary in thickness, cross-sectional shape, length and the material they are made of. The rods are held rigidly at one end, and weights placed on

the other to show how flexible they are. According to
Inhelder and Piaget's description, children at the concrete
operational stage have difficulty in controlling these
variables. Thus, asked to compare the flexibility of brass
and iron rods, children at this stage are likely to choose
rods of different length, cross-section, etc. If they
believe brass rods are more flexible they may even deliberately choose a brass rod that is longer and thinner than the
iron rod, and put a greater weight on its end. In contrast,
children at the formal operational stage will control the
extraneous variables by keeping them the same. In terms of
the hypothesis, 'If brass then more flexible (than iron)'
B F, this suggests that these children are aware of the
possibility that the four basic combinations B.F.,B.F,B.F. and
B.F may exist. However, this does *not* mean, as is required
in Wason's task, that they regard the possibility that B.F
might exist as crucial (a brass rod that is less flexible),
in the sense that they realise that the nearest they can come
to verifying the hypothesis is to try to demonstrate that B.F
cannot be true. Patently B.F can be true (if, compared to the
iron rod, the brass rod is sufficiently short and thick, etc.).
B.F is only important in the sense that it is recognized that
this, or any other combination, might occur through the effect
of extraneous variables. Thus these variables have to be
controlled, however, this being done, the aim is to *confirm*
the hypothesis, to see whether B.F is true. In terms of
Wason's task, this is equivalent to turning over the A and
the 4 cards, which is in fact what most subjects do.

The table below shows the frequency with which the
various combinations of cards were chosen on Wason's original
task, and in a number of subsequent studies in which the
same task was used with only minor variations (see Johnson-
Laird and Wason, 1970).

A and 4	52
A	42
A, 4 and 7	9
A and 7	5
Others	13

N = 128

These subjects were all university students, but, as can be
seen, only five subjects showed complete insight into the task,
with a further nine showing partial insight, but with most
choosing only the cards that verified the rule (A and four).

Subsequent studies (e.g., Evans and Lynch, 1973, and
Evans, 1972) suggest that some subjects may be using an even
more primitive strategy than the 'verification strategy'
suggested above, namely a 'matching strategy', whereby
subjects simply choose the cards mentioned in the original
rule. The tendency to use this strategy can be shown by

using negative components in the rule; for example, if the rule is 'If a card has a vowel on one side, then it does *not* have an even number on the other side', subjects using the verification strategy would choose 'vowel' and 'not even' (i.e., A and 7 if the cards were A, D, 4, 7), whereas subjects using a matching strategy would choose 'vowel' and 'even' (A and 4). The ringed numbers in each column of the table below (from a study by Manktelow and Evans (unpublished) involving 24 undergraduate students who were given rules similar to Wason's), suggests that a substantial number of subjects are using this latter strategy.

Frequencies of selection of various 'cards':

Rule	Logical Cases			
	True Antecedent	False Antecedent	True Consequent	False Consequent
If p then q	p: (23)	$\bar{p}$: 3	q: (15)	$\bar{q}$: 8
If p then not q	p: (24)	$\bar{p}$: 2	$\bar{q}$: 5	q: (18)
If not p then q	$\bar{p}$: 19	p: (7)	q: (17)	$\bar{q}$: 12
If not p then not q	$\bar{p}$: 20	$\bar{p}$: (5)	$\bar{q}$: 13	q: (16)

In Wason's original rule, the elements are fairly abstract and the relationship between them arbitrary. Later studies examined the effect of adopting a more realistic guise. Their findings suggest that the difficulty of the task is not so much due to its logical structure, *per se,* but to its content or mode of presentation. The first of these studies was by Wason and Shapiro (1971) who used the rule 'Every time I go to Manchester I travel by train' and presented subjects with four cards showing 'Manchester', 'Leeds', 'Train' and 'Car'. They found that ten out of 16 subjects correctly chose to turn over the cards 'Manchester' and 'car', compared to a control group of 16 subjects given the original rule, of whom only two chose the correct cards. Lunzer, Harrison and Davey (1972) found a similar effect using pictorial material; in their case, the rule was 'Every red lorry is full of coal', and subjects had to choose from 'full lorry', 'empty lorry', 'red' and 'yellow'. An even greater increase in facility was found by Johnson-Laird, Legrenzi and Legrenzi (1972) who used the rule 'If a letter is sealed, then it has a 50 lire stamp on it'. Subjects were asked to imagine they were post-office workers sorting letters; they were presented with the five envelopes shown below and instructed to 'select those envelopes that you definitely need to turn over to find out whether or not they violate the rule'. Subjects were given two such realistic tasks, and two symbolic tasks (in which the rule was similar to Wason's original rule, but envelopes were used rather than cards). The order of presentation of the tasks was varied in order to detect transfer effects from one condition to the other. The subjects were 24 undergraduate students, of whom

22 solved at least one of the tasks in the realistic condition, but only 7 managed to do so in the symbolic condition.

	Realistic Condition	Symbolic Condition
Both tasks correct	17	O
One task correct	5	7
Neither task correct	2	17
	—	—
	24	24

Johnson-Laird *et al.* report that there was a striking lack of
any direct transfer between the two types of task, and that
on subsequent questioning only two subjects acknowledged
that the underlying logical structure was the same. Lunzer
et al. (ibid.) make similar observations, which adds strong
support to the view that it is the content rather than the
logical structure which is crucial to solving the tasks.

A number of subsequent studies have tried to analyse
the effect of content in more detail. Van Duyne (1976)
distinguishes between realistic rules that can be interpreted as *necessarily* true (e.g. 'If it is raining then the
streets are wet') and rules that are *contingently* true
(e.g. 'If it is a tomato then it is red'). He found that
the necessarily true rules were significantly more difficult
($p < .01$). For any given rule subjects had to answer four
questions about whether, given one piece of information, it
was necessary to look for additional information to determine whether the rule is always true. In the case of 'If
it is a tomato then it is red', for example, the initial
information would be, in turn, 'It is a tomato', 'It is red',
'It is not a tomato', 'It is not red'. Answers were only
judged as correct if subjects adequately explained why
additional information was or was not necessary. The subjects were 22 undergraduates; the mean number of correct
answers (out of four) was 1.95 for the necessary condition
and 2.84 for the contingent condition.

The key to all tasks derived from Wason's original rule

is to look for instances that might violate the rule;
however, Van Duyne argues that subjects would tend to ignore
such instances when the rule is regarded as necessarily
true. thus while there may exist tomatoes that are not red,
there would seem to be no point in considering the condition
'It is raining - the streets are dry' since this is assumed
never to occur. Van Duyne goes on to suggest that the high
success rate on the rule used by Johnson-Laird *et al.* (1972)
can be explained by the non-arbitrariness of the rule, *and*
by the fact that it is liable to be violated, i.e., is
contingently true. It might also be added that whereas the
original wording used by Wason asks subjects to consider
whether the rule may be 'true or false', Johnson-Laird
et al. draw specific attention to the notion that the rule
may be violated.

Bracewell and Hidi (1974) conducted an experiment in
which they distinguished between the *terms* of the rule and
the *relationship* between the terms. Thus they used rules
(examples of which are shown below) in which the terms
involved either 'concrete' or 'abstract' material, and the
relationship was either 'natural' or 'arbitrary'. They also
examined the effect of reversing the order of the terms by,
for example, rewriting the rule 'Every time I go to Ottawa
I travel by car' to 'I travel by car every time I go to
Ottawa'. (Notice here that the meaning of the rule has not
changed.)

	Concrete Material	Abstract Material
Natural Relationship	Every time I go to Ottawa I travel by car	Every time I go to J I travel by 2
Arbitrary Relationship	Every time Ottawa is on one side car is on the other side	Every time J is on one side 2 is on the other side

Altogether, 96 graduate students participated in the
study. For each task subjects were awarded 3 points for
a 'complete insight' response (selecting the true antecedent
and the false consequent), 2 points for a 'partial insight'
response (selecting the true antecedent and the false
consequent and the true consequent) and 1 point for any other
response. The table below shows the mean scores for each of
the eight conditions.

	terms in orthodox order		terms in reverse order	
	Concrete Material	Abstract Material	Concrete Material	Abstract Material
Natural Relationship	2.67	1.75	1.42	1.58
Arbitrary Relationship	1.42	1.17	1.00	1.17

As can be seen from the above table, the natural relationship appears to be easier than the arbitrary relationship, the orthodox order appears to be easier than the reverse order, and the concrete material appears to be easier than the arbitrary material (but *only* for the orthodox order). These differences were confirmed by an analysis of variance.

As far as the relationship factor is concerned, Bracewell and Hidi suggest that 'the arbitrary relationship conditions imposed more of a cognitive load on the subject than the natural relationship conditions' (p.486). In particular, they argue that this is the case when it comes to realizing that the relationships are non-reversible, that 'if p then q' is not the same as 'if q then p': for arbitrary relationships subjects may have to remember this explicitly, whereas this non-reversibility is inherent in natural relationships.

It is not at all obvious that the rule 'Every time Ottawa is on one side, car is on the other side' is different in meaning from 'Every time car is on one side, Ottawa is on the other side'. On the other hand, the knowledge evoked by the rule 'Every time I go to Ottawa, I travel by car' makes its reversal, 'Every time I travel by car I go to Ottawa' sound distinctly odd: from everyday experience, it is known that cars are used for going to many different places.

It is possible to take Bracewell and Hidi's argument a step further, to construct the same argument as was used in the case of O'Brien's studies. Thus not only does a natural relationship like 'If I go to Ottawa, I travel by car' inherently carry the information that it is non-reversible, but it also readily evokes each of the combinations 'Ottawa, car', 'not Ottawa, car', 'not Ottawa, not car' and 'Ottawa, not car', together with the realization that the last of these contradicts the relationship. Put another way, the relationship evokes a familiar or plausible 'frame' from which it is possible, given the value of one term (e.g., 'not car'), to read off the values of the other term that make the relationship true ('not Ottawa') or false ('Ottawa').

	I travel by car	I do not travel by car
I go to Ottawa	true	false
I do not go to Ottawa	true	true

The fact that Bracewell and Hidi found no significant effect on task difficulty for the materials factor alone, underlines the point made with respect to O'Brien's studies, that the use of concrete, or familiar, terms does not necessarily make the relationships as a whole meaningful. However, a study by Gilhooly and Falconer (1974) has shown that concrete terms can make the task easier, and in the present study the concrete materials (Ottawa, car) were easier than the abstract materials (J, 2) when the terms were presented in the orthodox order. Bracewell and Hidi argue that this might be because the particular concrete terms they used express a goals-means relationship (with cities as destinations or goals, and the transportation as means to the goals) which the abstract terms do not; furthermore the orthodox order of presentation may correspond to the termporal order of thinking about goals and means, with the result that when the second order is adopted, the relationship is obscured, and the advantage of using concrete terms lost.

Van Duyne (1976) refers briefly to a study in which he used rules in four different conditions, which can be described as follows (using Bracewell and Hidi's terminology):

(1) *Abstract* terms, *arbitrary* rule (e.g., 'If a vowel is on one side, then there is an even number on the other');
(2) *Concrete* terms, *arbitrary* rule (e.g., 'If there is L.B.Mill on one side of the envelope then there is PRINTED MATTER REDUCED RATE on the other side', with real envelopes as stimulus material),
(3) *Concrete* terms, *natural* rule (simulated) (e.g., 'If there is PRINTED PAPER REDUCED RATE on one side of the envelope then it must be left open', with white cards representing real envelopes);
(4) *Concrete* terms, *natural* rule (e.g., 'If there is PRINTED PAPER REDUCED RATE on one side of the envelope then it must be left open', with real envelopes as stimulus material).

Van Duyne reports that the percentages of correct solutions (for, presumably, adult subjects) were (1) 19 per cent, (2) 49 per cent, (3) 87 per cent, (4) 98 per cent. The large difference between condition (2) and conditions (3) and (4) supports the view that the critical influence on logical reasoning is the degree to which the rule as a whole is meaningful - in the sense of being familiar or plausible.

A number of studies have examined ways of improving performance on the standard (abstract) Wason task. Roth (1979) makes the point that to solve this task subjects need to consider what symbols might occur on the back of *each* of the four cards. Subjects who are operating under either a 'verification bias' or a 'matching bias' appear not to be doing this: given, say, the rule 'If vowel then even' (If P then Q) and cards showing the symbols A, D, 4, and 7 (P, $\bar{P}$, Q, $\bar{Q}$), these subjects seem impulsively to focus on A and 4 only (P and Q). Roth argues that this may arise because the task appears deceptively simple, rather than because subjects are unable to assess the significance of what each of the cards may reveal. To test this, Roth presented a group of subjects with a modified four-card array, in which the P card was replaced by another $\bar{Q}$ card, on the assumption that this would reduce the tendency to respond impulsively, and would, in particular, focus attention on the possible significance of $\bar{Q}$. A second group of subjects was given a standard array, and both groups were then given a transfer task involving another standard array. There were 28 subjects in each group, all undergraduates. The table below suggests that the use of the modified array had a substantial effect on subjects' understanding of the standard transfer task, at least to the extent of inducing 'partial insight'. However, Roth does make the point that an attempt to make these subjects gain 'complete insight', by forcing them subsequently to choose two rather than three cards, failed: they were as likely to reject $\bar{Q}$ as Q. Also Roth did not administer a delayed post-test to determine whether the gains made were permanent.

		Response-type			
		Complete Insight	Partial Insight	Bi-Conditional	No Insight
		P$\bar{Q}$	PQ$\bar{Q}$	P$\bar{P}$Q$\bar{Q}$	Other
Transfer Task	modified array group	6	10	2	10
	standard array group	7	3	0	18

An extensive study by Abbott (1974) considered the effect of training on performance of Wason's task, and also investigated the effect of a more realistic content and of a modified array (similar to Roth's) on task difficulty. The sample consisted of four groups of subjects, all of above average intelligence, whose ages and number are shown below.

Group A: 21 children aged 7:09 to 8:09
Group B: 21 children aged 10:06 to 11:06
Group C: 21 adolescents aged 13:00 to 14:00
Group D: 24 mature students.

Five variants of Wason's task, which differed in content and in presentation but which were of the same logical form, were given as pre- and post-tests (Task 1, 2, 3, 4, 6). In addition a far more complex, '3-dimensional' version of the task was used (Task 5). The rules for each task are shown in abbreviated form below. Task 1 is the same as an early task used by Wason, which involves cards with a circle on one side and a triangle on the other; four of these are presented, showing a red triangle, a blue triangle, a red circle and a blue circle (i.e. P, $\bar{P}$, $\bar{Q}$ and Q, for the rule P→Q). Similar cards are used in Tasks 3 and 6, except that for Task 6 the P and the P card are replaced by a second Q and a second Q card. Task 2 involves small film canisters, representing tins, some of which are open and some of which are partly hidden. Task 4 uses toy trumpets some of which have been flattened and some of which are hidden. The more complex Task 5 uses four triangular prisms, each with a square on one rectangular face, a circle on another and a triangle on the third, but each with only one rectangular face visible.

Rule	Materials
Task 1 : Red △ → Blue ○	Cards
Task 2 : Tin with Black Band → Full	Canisters
Task 3 : Red □ or Yellow △ → White ○	Cards
Task 4 : Flattened Trumpet → Honks	Trumpets
Task 5 : Red □ or Blue ○ → Yellow △	Prisms
Task 6 : Orange Card → ○	Cards

After the pre-tests subjects were given extensive training involving a further seven variants of the task. For each of the tasks subjects were asked, with respect to the categories of card P, $\bar{P}$, Q and $\bar{Q}$,

1. to state what could be on the other side;
2. to state what effect this would have on the rule;
3. to state whether the card should be turned,
4. to turn the card, state what effect the revealed evidence has on the rule, and modify the selection if necessary.

After each response subjects were told the correct response and given a reason for it.

On the pre-test, for Tasks 1, 2, 3, 4 and 6, there was no significant difference in performance between the four groups of subjects. However, compared to Task 1, Task 4 was significantly easier (when a strict criterion of marking was used: selection of P and $\bar{Q}$), and Task 3 was significantly harder (using a less strict criterion of marking: P and $\bar{Q}$, or P, Q and $\bar{Q}$). Rather surprisingly, Tasks 2 and 6 were of similar difficulty to Task 1. The relative ease of Task 4 may have been due not only to its more realistic content, but more specifically to the fact that the *im*plausible nature of the rule may have focused attention on seeking instances that would violate the rule.

On the post-test (i.e., after training), performance on Tasks 1, 2, 3, 4 and 6 improved substantially, as can be seen from the table below. However, there was no longer any significant difference between these five tasks.

Number of correct responses (P and $\bar{Q}$) for all subjects (N=87)

Task	1	2	3	4	6
pre-test	4	4	0	12	6
post-test	45	38	39	37	39

Despite the over-all improvement, subjects in the youngest group (A) performed significantly worse than the other subjects. On the whole, the performance of subjects in group B was similar to the two oldest groups, with, for example, 10, 11 and 13 subjects in group B, C and D respectively solving at least three of the five tasks.

The post-test results suggest that training produced a substantial improvement in subjects' understanding of the standard Wason task. However, after having examined the results of Task 5, Abbott's conclusions are more cautious. Task 5 proved to be extremely difficult, with no subjects showing any insight on the pre-test. The table below shows that performance improved on the post-test, particularly for the older subjects (groups C and D); however, only five subjects showed complete insight into the task, and of the younger subjects (groups A and B), only two gave responses that, according to Abbott, show any insight at all. Abbott concludes from this that the effect of training on the younger subjects was to induce algorithmic learning, i.e., the learning of a rule, rather than genuine understanding. Moreover, these subjects seemed satisfied with algorithmic learning despite the efforts made by the experimenter to induce insight.

Post-test Responses on Task 5 :	No Insight	Some Insight
Group A	21	0
B	19	2
C	12	9
D	15	9

Implications for teaching

It is very clear from these studies that logical thinking in
a familiar context is different from, and much easier than,
formal logic, which can cause difficulties even for highly
intelligent adults. (Incidentally, the studies also show that
the tasks described by Piaget as formal do not require an
understanding of formal logic.) In mathematics similar
differences are likely to occur; this means it should be
possible to help children cope with certain proof activities,
for example, by choosing contexts that are sufficiently
meaningful or familiar, but few children are likely to achieve
a formal understanding of the underlying processes.

Training studies have shown that performance on logical
tasks can be substantially improved, particularly when it
comes to accepting that it is not always possible to make
unambiguous deductions; however, with younger children
especially, the improvement is likely to result from rule
learning rather than from a real gain in insight into the
tasks' logical structure. This, and the fact that even
adults were found to have difficulty recognizing the common
structure of familiar and abstract logical tasks, exposes a
dilemma that may well apply to many areas of mathematics:
choosing a more familiar context, while making a task easier
to understand, may at the same time obscure the task's underlying structure and thereby reduce the likelihood of transfer
to more abstract versions of the task.

References for Section 2 'Memory'

BOWER, G.H. and CLARK, M.C. (1969): 'Narrative stories as mediators for serial learning', *Psychometric Science,* 14, 181-2.

BUGELSKI, B.R. (1968): 'Images as a mediator in one trial paired associate learning', *J Exp Psych,* 77, 328-34.

CASE, R. (1974): 'Structures and strictures, some functional limitations on the course of cognitive growth', *Cog Psych,* 6, 544-73.

CASE, R (1975): 'Gearing the demands of instruction to the developmental capacities of the learner', *Rev Ed Res,* 45, 59-87.

CASE, R. (1978): 'A developmentally based theory and technology of instruction'. *Rev Ed Res;* 48, 439-63.

COLLIS, K.F. (1975(a)): *A study of Concrete and Formal Operations in School Mathematics: A Piagetian Viewpoint.* Melbourne: Australian Council of Educational Research.

COLLIS, K.F. (1975(b)): *The Development of Formal Reasoning.* New South Wales: University of Newcastle.

CRAIK, F.I.M. (1970): 'The fate of primary memory items in free recall', *J Verbal Learning and Verbal Behavior,* 9, 143-8.

CRAIK, F.I.M. and LOCKHART, R.S. (1972): 'Levels of processing: a framework for memory research', *J Verbal Learning and Verbal Behavior,* 11, 671-84.

CRAIK, F.I.M. and MASANI, P.A. (1969): 'Age and intelligence differences in coding and retrieval of word lists', *B J Psych,* 60, 315-19.

CRAIK, F.I.M. and TULVING, E. (1975): 'Depth of processing and the retention of words in episodic memory', *J Exp Psych: General,* 104(3), 268-94.

CRAIK, F.I.M. and WATKINS, M.J. (1973): 'The role of rehearsal in short-term memory', *J Verbal Learning and Verbal Behavior.* 12, 599-607.

CRANNELL, C.W. and PARISH, J.M. (1957): 'A comparison of immediate memory span for digits, letters and words', *J Psych,* 44, 319-27.

FLAVELL, J.H., BEACH, D.R. and CHINSKY, J.M. (1966): 'Spontaneous verbal rehearsal in a memory task as a function of age', *Child Dev,* 37, 283-99.

GRONINGER, K.D. (1971): 'Mnemonic imagery and forgetting', *Psychometric Science,* 23, 161-3.

GRUNEBERG, M.M. and MONKS, J. (1974): 'Feeling of knowing and cued recall', *Acta Psych,* 38, 257-65.

GRUNEBERG, M.M. (1978): 'The feeling of knowing, memory blocks and memory aids', In: GRUNEBERG, M.M. and MORRIS, P. (Eds): *Aspects of Memory.* London: Methuen.

HAGEN, J.W. and KINGSLEY, P.R. (1968): 'Labelling effects in short-term memory', *Child Dev,* 39, 113-21.

HALFORD, G.S. (1978): 'Towards a working model of Piaget's stages'. In: KEATS, J.A., COLLIS, K.F. and HALFORD, G.S. (Eds): *Cognitive Development.* Chichester: Wiley.

HARTLEY, J.R. (1980). Using the Computer to Study and Assist the Learning of Mathematics. Paper presented to the BSPLM, University of Nottingham, January 1980.

HITCH, G.J. (1978): 'The role of short-term working memory in mental arithmetic', *Cog Psych,* 10, 302-23.

HUNTER, I.M.L. (1964): *Memory.* Harmondsworth: Penguin Books.

HUTTENLOCHER, J. and BURKE, D. (1976): 'Why does memory increase with age?', *Cog Psych,* 8, 1-31.

KEENEY, R.J., CANNIZZO, S.R. and FLAVELL, J.H. (1967): 'Spontaneous and induced verbal rehearsal in a recall task', *Child Dev,* 38, 953-66.

KINGSLEY, P.R. and HAGEN, J.W. (1969): 'Induced versus spontaneous rehearsal in short-term memory in nursery school children', *Dev Psych,* 1, 40-6.

KREUTZER, M.A., LEONARD, C. and FLAVELL, J.H. (1975): *An Interview Study of Children's Knowledge about Memory.* Monograph of the Society for Research in Child Development, 40 (1, serial number 159).

LAWSON, A.E. (1976):'M-space: is it a constraint on conservation reasoning ability?' *J Exp Child Psych,* 22, 40-9.

LINDSAY, P.H. and NORMAN, D.A. (1972): *Human Information Processing: An Introduction to Psychology.* New York: Academic Press.

McLAUGHLIN, G.H. (1963): 'Psycho-logic: a possible alternative to Piaget's formulation', *B J Ed Psych,* 33, 61-7.

MILLER, G.A. (1956): 'The magical number seven, plus or minus two: some limits on our capacity for processing information',

Psych Rev, 63, 81-97.

PASCUAL-LEONE, J. and SMITH, J. (1969): 'The encoding and decoding of symbols by children: a new experimental paradigm and neo-Piagetian model', *J Exp Child Psych,* 8, 328-55.

PIAGET, J. and INHELDER, B (1969): *The Psychology of the Child.* London: Routledge.

RABBITT, P.M. (1968): 'Channel-capacity, intelligibility and immediate memory', *Quarterly J Exp Psych,* 20, 241-8.

SHULMAN, H.G. (1971): 'Similarity effects in short-term memory', *Psych Bull,* 75, 399-415.

TRESSELT, M.E. and MAYZNER, M.S. (1960): 'A study of incidental learning', *J Psych,* 50, 339-47.

WITKIN, H.A., DYK, R.B., FATERSON, H.F., GOODENOUGH, D.R. and KARP, S.A. (1962): *Psychological Differentiation.* New York: Wiley.

WOODWARD, A.E., BJORK, R.A. and JONGEWARD, R.H. (1973): 'Recall and recognition as a function of primary rehearsal', *J Verbal Learning and Verbal Behavior,* 12, 608-17.

References for Section 3 'Logic'

ABBOTT, M.A. (1974). The effect of training and developmental level upon the performance of an inferential task. M.Phil. thesis, University of Nottingham.

BRACEWELL, R.J. and HIDI, S.E. (1974): 'The solution of an inferential problem as a function of stimulus materials', *Quarterly J Exp Psych,* 26, 480-8.

EVANS, J. St.B.T. and LYNCH, J.S. (1973): 'Matching bias in the selection task', *B J Psych,* 64, 391-7.

EVANS, J.St.B.T. (1972): 'Interpretation and matching bias in a reasoning task', *Quarterly J Exp Psych,* 24, 193-9.

GILHOOLEY, K.J. and FALCONER, N.A. (1974): 'Concrete and abstract terms and relations in testing a rule', *Quarterly J Exp Psych,* 26 (3), 355-9.

HADAR, N. (1978): 'Children's Conditional Reasoning Part III: a design for research on children's learning of conditional reasoning and research findings', *Ed Stud Math,* 9, 115-40.

INHELDER, B. and PIAGET, J. (1958): *The Growth of Logical Thinking from Childhood to Adolescence.* London: Routledge.

JOHNSON-LAIRD, P.N. and WASON, P.C. (1970): 'A theoretical analysis of insight into a reasoning task', *Cog Psych*, 1, 134-48.

JOHNSON-LAIRD, P.N., LEGRENZI, P. and LEGRENZI, M.S. (1972): 'Reasoning and a sense of reality', *B J Psych*, 63, 395-400.

KNIFONG, J.D. (1974): 'Logical abilities of young children - two styles of approach'. *Child Dev*, 45, 78-83.

LUNZER, E.A., HARRISON, C. and DAVEY, M (1972): 'The four-card problem and the generality of formal reasoning', *Quarterly J Exp Psych*, 24, 326-39.

MANKTELOW, K.I. and EVANS, J.St.B.T.: Facilitation of Reasoning by Realism: Effect or Non-effect? Unpublished paper: Plymouth Polytechnic.

O'BRIEN, T.C., SHAPIRO, B.J. and REALI, N.C. (1971): 'Logical thinking - language and context', *Ed Stud Math*, 4, 201-19.

O'BRIEN, T.C. (1972): 'Logical thinking in adolescents', *Ed Stud Math*, 4, 401-28.

O'BRIEN, T.C. (1973): 'Logical thinking in college students', *Ed Stud Math*, 5, 71-80.

POPPER, K.R. (1959): *The Logic of Scientific Discovery*. London: Hutchinson.

ROTH, E.M. (1979): 'Facilitating insight in a reasoning task', *B J Psych*, 70, 265-71.

VAN DUYNE, P.C. (1976): 'Necessity and contingency in reasoning', *Acta Psych*, 40, 85-101.

WASON, P.C. (1966): 'Reasoning'. In: FOSS, B. (Ed), *New Horizons in Psychology*, Harmondsworth: Penguin Books.

WASON, P.C. and SHAPIRO, D. (1971): 'Natural and contrived experience in a reasoning task', *Quarterly J Exp Psych*, 23, 63-71.

Chapter Three
Stages of General Intellectual Development

1. INTRODUCTION

The material of this chapter relates to the teaching of
mathematics in three distinct ways. First, if it is possible
to recognize general stages of intellectual development, each
representing a characteristic mode of reasoning and a range
of types of task with which children are capable of dealing,
then this information is helpful to the teacher in designing
teaching material and in responding to questions and to performances by his pupils. It enlarges the range of concepts
by which he can interpret their actions and the responses.
The work of Collis and of the Chelsea Project, *Concepts in
Secondary Mathematics and Science,* as well as the original
work of Piaget himself, all offer insights into the way in
which children's general development of ways of thinking
relates to their performance in mathematics. In this respect
the work of this chapter constitutes generalizations from the
particular detailed descriptions of children's understanding
of particular mathematical topics which appears in Chapter
Seven. The value of this work still stands even if the search
for mathematical stages going across different topics is not
very successful, and we have to be content with seeing stage
developments *within* topics.

Secondly, if there indeed exist general ways of functioning, characteristic of particular stages, for example if there
exist certain kinds of logical reasoning which children are

capable of, and these are the mental tools which they use in their learning across all topics, then there would be some interest in the possiblity of achieving a general acceleration of this development which might give children more powerful learning tools at an earlier age. Unfortunately the training studies which have been performed and which are described in the middle part of this chapter show success in the improvement of logical reasoning only in limited areas. There may well be general educational interventions which might produce more all-round development, but these are likely to consist not of relatively short and simple training procedures, but rather the general encouragement of reflection and reasoning throughout the curriculum.

Thirdly, if children's reasoning capacities are relatively general, and also rather stable and slow to develop, then it becomes important to consider whether there are parts of the curriculum which can only be satisfactorily learnt once a particular stage has been reached. Thus the theory would provide a rough guide for curriculum design. Work by Shayer adopting this point of view in relation to the science curriculum is described at the end of this chapter. Other workers have tended to turn their attention rather to the study of learning processes in particular topics or even particular tasks, partly because they feel that the coherence of the stages is not sufficient, and partly because they feel it more valuable to attempt to devise ways of teaching a particular conceptual structure than to decide in advance that it is inaccessible.

2. THE PIAGETIAN STAGES

The classical Piagetian position is that the modes of reasoning which children use pass through a number of broad stages: a *sensori-motor* stage up to 18 months, then a *preoperational* stage, roughly up to the age of six; a *concrete operational* stage lasting from about six to about eleven, and a *formal operational* stage, covering the ages of approximately eleven to sixteen. During the preoperational stage children first become aware of the permanence of objects, that is, that objects have an independent existence, independent of whether they are being observed by the child; later they become able to represent objects or events by symbols, that is, by words or drawings. During the concrete operational phase the child makes mental representations of the *actions* he performs in such a way as to enable him to imagine the reversal of an action and the compensation of one action by another. This enables him, among other things, to recognize the concept of number as being a property of a set of objects which is unchanged when the members of the set are spread out or closed up in relation to each other. He can recognize that the spreading operation can be reversed to return the elements to their original positions; he can also recognize that the greater space occupied by the spread-out set is compensated by the fact that the objects fill this space in a sparser way. He can also make the one-one correspondence needed to match two sets. This elementary type of logic is also needed to construct the concepts of quantity and length, area and volume, though these represent situations of increasing difficulty to which the logic has to be applied. Also at this stage the child is able to make a series out of a set of objects by ordering them, as when a set of sticks is ordered by length. Thus, in Piaget's words, 'The concrete operational thinker collects results, classifies and orders them, and establishes correspondences' (Inhelder and Piaget, 1958). By contrast, the formal operational thinker is able to reason by forming and testing hypotheses; in Piaget's words, he will view 'a given correspondence as the result of one of several possible combinations, and this leads him to verify his hypotheses by observing their consequences.' A particular characteristic mentioned by Piaget as determinant of the formal stage is the ability to separate variables in an experiment; for example when four bottles of colourless liquid are given and the subject asked to find how to make a suitable mixture of them which will produce a colour with a given indicator solution, the concrete thinker tests combinations of the four bottles at random, whereas the formal thinker uses a scheme by which all possible combinations are tested in turn. Similarly, given a simple pendulum formed by strings of various lengths and with weights of different magnitudes, and asked to find what determines the frequency of the swing, the formal thinker will have an 'anticipatory scheme' which ensures that each variable is tested while the others are kept constant, whereas the concrete thinker is unable to achieve this separation of the variables and changes two or more variables at once.

Shayer (1979) makes the point that Piagetian theory has three 'tiers'. The lowest of these consists of the descriptions of children's behaviour under various experimental tasks. The second **tier** is Piaget's classification of these behaviours into seemingly ordered and coherent developmental stages, whereas the third tier attempts to model these stages by means of logico-mathematical structures.

The first tier is uncontroversial, at least as far as some of the more classical Piagetian experiments, such as his number conservation tasks, are concerned; these tasks have been thoroughly replicated and by and large the reported behaviours have been confirmed. Rather it is the interpretation of these behaviours, both in terms of stages and in terms of theoretical logico-mathematical structure that is in dispute. Given any two propositions p and q, Piaget argues that while children at the concrete stage can employ only a limited logic of classification, correspondence and ordering, at the formal operational stage they construct and test possible relationships between propositions using the whole structure whose elements are all 16 combinations

53

of the binary operations $p.q$ (p and q), $\bar{p}.q$ (not p and q), $p.\bar{q}$ and $\bar{p}.\bar{q}$ (see Boyle, 1969).

Having constructed a model based on propositional calculus to describe the structure of formal operational thought, Piaget seems to have taken the further step of arguing that children at this stage actually have access to the model itself. This is analogous to arguing that a baby who has learnt to crawl, and whose movements might be described by a group of displacements, understands the structure of a group. Even the demonstration of the high degree of dependence on context (see Chapter Two and section 4a below) does not mean that the model should be rejected out of hand; it may still be useful as a descriptive and heuristic device. An examination of the tasks Piaget used to identify formal thinking in fact suggests that the propositions that children were able to test (e.g., $p{\rightarrow}q$ expressed as 'if a heavier weight then more swings' in the pendulum task) enjoyed precisely the same kind of contextual support as Johnson-Laird's stamps-envelopes problem.

The experiments on which this theory is based show quite striking differences between younger and older subjects. Many years of world-wide experiment have shown that the order of acquisition of the concepts embodied in the various tasks used to define the concrete operational phase is constant across cultures, though the actual ages of acquisition may be shifted earlier or later as a whole. The degree of coherence of the stages is a matter of some dispute, and will be discussed later in this chapter. (A useful introduction to Piagetian theory can be found in *The Psychology of the Child*', Piaget and Inhelder, 1969.)

3. THE STAGES AND MATHEMATICAL DEVELOPMENT

Collis, Halford and Case

Collis has attempted to relate his mathematical tasks to the characteristics of performance shown in the quantitative scientific tasks studied by Inhelder and Piaget (1958). Thus, he says, at stage 2A (early concrete operations, approximately seven to nine years) the child can perform operations on immediately observable physical phenomena; in Piaget's experiment on the beam balance the child is able to add the weights he is using and compare the distances, but is not able to coordinate weight and distance in anything more than a very rough and intuitive way. Similarly, in the Piaget experiment on the projection of shadows, the child at this stage knows that the size of the shadow depends on the size of the object but this knowledge goes no further. Collis' experiments with mathematical tasks suggest that at this stage both the elements and the operations of arithmetic must be related directly to physically-available elements and operations, for example, the child can calculate that

3 + 5 = 8 by imagining sets of three and five objects put together and counted. At this stage, an expression such as 3 + 8 + 5 has no meaning for the child since he cannot conceive of 3 + 8 as representing a new number until he has actually performed the operation and obtained the result 11; that is, in Collis' terms, until he has actually 'closed' the operation. At this stage also, subtraction can have a meaning, but only in physical terms, that is, that what is put down can be taken up.

At stage 2B (late concrete operations, age 10 - 12), in Piaget's experiment on the beam balance children recognize the qualitative correspondence between weight and distance, expressed for example as 'the heavier it is, the closer to the middle'. In the shadows experiment the qualitative relation between shadow size and distance is also clearly formulated and there are in fact the beginnings of attempts at metrical quantification, but only in cases where there are constant additive differences in the sizes given. In Collis' mathematical work, at this stage the child recognises the uniqueness of the result of a single operation with small numbers and thus can cope with an item requiring him to compare 3 + 8 + 5 with 5 + 8 + 2, or 479 x 231 with 456 x 231; in the case of the items with larger numbers, only one operation can be handled. With regard to inverses, although an equation such as y + 4 = 7 can be solved easily, and although it may be recognized that this answer is obtained by subtracting 4 from 7, this does not amount to a general recognition that subtraction can always be used to undo an addition.

Stage 3A (early formal operations, 13 - 15 years), is called *concrete generalization* by Collis. At this stage, in the beam balance experiment, the child notices that to restore the beam from the inclined position to the level a point at greater distance from the fulcrum needs to travel through a greater height, so that a weight placed at a greater distance has to travel farther than one closer to the fulcrum; but the translation of this observation into an explanation of the weight x distance law in terms of equal amount of work does not in general appear at this stage. In the shadows experiment an inverse metrical proportionality between distances and diameters of the rings appears but is not generalized to all possible cases. The proportion is found in one or two instances only. The child tends to be satisfied when he has verified his hypothesis on a single case; he does not yet look for a *general* law. In Collis' mathematical items he observes that at this level children work on the basis of concrete generalizations where a few specific instances satisfy them of the reliability of a rule, but the rules can be applied only in cases where the operations are recognized as being performable to give a unique outcome; for example, the child would be able to determine that $382 \times 743 = 672 \times 743$ but would not necessarily be

$\qquad\qquad\quad$ 382 $\qquad\qquad$ 672

able to understand and use meaningfully the generalization
$ma/m = na/n$. At this stage the notion of inverse is general,
that is, it is known that an expression like $y + 4$ can be
reduced to y by subtracting 4.

Stage 3B (late formal operations, 15+ years). In
Piaget's balance experiment, at this stage the law $w/w^l = 1/l^l$
is stated in full generality, and in the shadow experiment
again the law is fully stated in metrical terms; from the
start, the formulations are dependent on a hypothesis that
is both explanatory and general, and the hypothesis no longer
deals only with the divergent light rays, but includes a
conception of the cone of light itself. In the mathematical
items, Collis' subjects come to the problem with a set of
abstract hypotheses to test, and they are no longer satisfied
that one or two specific demonstrations are a sufficient
basis on which to generalize. They can conceive of a defined
operation such as $a*b = a + 2b$ and solve such problems as
'Does $a*b = b*a$', whereas younger pupils cannot overcome their
their tendency to regard an operation such as $*$ as being
necessarily one of the known operations $+$, $-$, $\times$ or $\div$. At
this stage also, a general notion of inverse can be handled,
for example an item such as 'if $(por)oq = (aob)oq$, then
$por = aob$ (Collis, 1975).

The dimensions of development identified here include:-
(i)   centration on one operation (stage 2A) leading to ability
     to co-ordinate two operations (2A/3B), and then to co-
     ordination of more than two (stage 3B) (in proportion,
     the relations among *three* elements have to be co-
     ordinated simultaneously);
(ii) general idea of dependence (stage 2A), leads to qualit-
     ative correspondence,(if A goes up, B goes down) (2B),
     and then to quantitative relationship (3A);
(iii)a law recognized over a limited range or tested on a few
     cases (3A), leads to a fully generalized law, with
     explanation and/or justification sought (3B).

Collis introduced the term 'acceptance of lack of
closure' to describe the situation when an item such as
$576 + 495 = (596 + 382) + (495 - \square)$ is solved by recognizing
that the $+382$ in the first bracket must be compensated by
subtracting 382 in the second bracket; the 'lack of closure'
which is being handled in this case is the acceptance of the
contents of the brackets as representing numbers and therefore being meaningful without having to calculate the result
to give a visible number. Lunzer broadened this concept in
an experiment where children were required to find the operations performed by four buttons on a piece of apparatus containing also a light. At a certain stage in attempting to
solve this problem, it was necessary for success to
identify a button as being either an 'off' or a 'change'

button, but without at that stage knowing which of these it was to be. It was eleven-year-old children who were the first to be able to use an ambiguous label such as 'off or change' in the solution of this problem (Lunzer, 1973). In this case the 'lack of closure' is the ability to handle the partial, ambiguous knowledge.

Halford (1978) has attempted a more sweeping categorization of the difference between concrete and formal thinking in terms of mathematical items. He suggests that the pre-operational stage is that of unary operations or binary relations. The concrete stage is that of compositions of binary operations. By 'composition', it is meant that two binary operations have to be co-ordinated and cannot be closed in sequence, for example, the item $(2 \square 2) \square 2 = 6$ is classed as concrete since it contains one unknown operation, used twice, whereas the item $(4 \square 2) \square 3 = 2$, where *different* operations must be used to fill the two boxes, is formal. On Halford's sample the first item was done correctly by 65 per cent of pupils aged nine, the second by 65 per cent only at age 12. In solving the first of these two items, the pupil has only to try the four operations, $+$, $-$, $\times$, $\div$ in succession in each of the two boxes, whereas the solution of the second requires in principle each operation to be tried in the first box together with all possible operations in the second. It is thus yet another kind of 'acceptance of lack of closure'; a temporary choice for the first operation must be made while trials are performed on the second. Though Halford's classifications predict quite well the order of difficulty of the items as shown by testing, it is probably chasing a chimera to attempt to explain all intellectual development by a single dimension.

As has been discussed more fully in Chapter Two, a number of psychologists, in particular Pascual-Leone and Case, have recently put forward theories described as 'neo-Piagetian' which attempt to explain the body of Piagetian empirical results in terms of variations with age in the number of items which can be stored in an individual's working memory. Collis has also related his work to such theories in an experiment in which he showed that his supposed concrete level items were soluble by groups of subjects having short-term memory spans between five and six digits, while his formal items were solved by subjects who had as a group memory spans between six and seven digits (Collis, 1975). (For later work relating memory span to mathematical tasks see Romberg and Collis, 1980.) In Pascual-Leone's theory, the size of working memory increases from $e + 1$ at the age of three at the rate of one unit per two years up to $e + 7$ at age 15 (e represents an additional capacity to hold in mind an 'executive scheme', that is, the awareness of what task has to be performed). In Case's theory, the size of memory is constant at four units, but the type of unit which can be stored increases in complexity between the four main Piagetian stages (sensori-motor, preoperational, concrete and formal). Also in Case's theory, it is assumed that operations which have become automatic do not take up memory space, whereas those which still require thought for their direction do. These theories are comparatively new and both contain serious problems regarding what is to be counted as a unit for storage in the case of a particular task. It seems clear that this cannot be determined in the abstract but requires knowledge of the ways in which a particular child approaches the task, to recognize what he treats as an element in his thinking. However, the observation of children in problem-solving situations makes it very clear that the general notion of the number of items of information which have to be held in memory and co-ordinated is a most important determining factor in the difficulty of a problem.

The Chelsea CSMS Project

In this section it is proposed to compare children's performances on several of the CSMS tests, in order to see the extent to which it is possible to find (postulate) *general criteria* for explaining children's mathematical difficulties. To simplify the analysis, this will be done by examining the facility of selected items for the 14-year-old samples. This involves a host of assumptions, not least that the correlations between the tests, and between the selected items within a test, are sufficiently high for the items to be regarded as measuring the 'same thing'; for a discussion of these assumptions and the evidence that might support them, see Hart (1980).

The item-facilities can be examined in two ways: *horizontally*, by which is meant comparing items with the same facility from different tests, and *vertically*, which involves comparing the difference in facility between a pair of items from one test with a pair from another. Though horizontal comparisons sound simpler, they are more difficult to make sense of, since there is any number of possible reasons why an item has a particular (absolute) facility; on the other hand, the difference in facility between a pair of items may be due to a single, and easily identifiable change in the items, and it is therefore proposed to discuss vertical comparisons first.

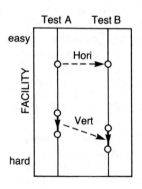

Vertical Comparisons. There are a number of ways in which vertical differences can be described. Perhaps the most

global, undifferential descriptor is to talk about the change
in 'complexity' of an item. Thus, for example, while over
90 per cent of the children could plot the points (2,5),
(3,7) and (5,11), less than 80 per cent could plot (1½,4); on
the eels questions on the ratio test, over 80 per cent could
double, but when the multiplier was changed to x3, x1½ and
x5/3 the facility was reduced by about 10 per cent, another
25 per cent and another 20 per cent respectively; changing
the context, from eels, to a geometrical enlargement reduced
the facility by about 30 per cent (for x1½), as did the
change from a single point to a flags, for reflections. In
each of these cases the harder situations involve an increase
in the complexity of the element such as from 3
to 5/3, from one dimension (eels) to two, from a one-point
to a two-point figure (point to flag). This is seen most
strongly in regard to numbers. Thus changes from small
positive whole numbers to large whole numbers, to negative
whole numbers, unknowns, fractions or decimals may all have
a substantial effect on facility (apart from the first pair
of items, facilities are for 14-year-olds):

		Facility	Change in Facility
small numbers to large numbers:	write a story 9 x 3 for 84 x 28	45% 31%	14% (11yr olds)
whole numbers to fractions:	plot (2,5)(3,7)(5,11) (1½,4)	91% 77%	14%
	volume of drawn cuboid of dimensions:		
	3 x 2 x 2 2 x 2½ x 2½	68% 28%	40%
	eels: x 3 x 1½ x 5/3	75% 50% 30%	25% 20%
	area of drawn rectangle of dimensions: 6 x 10 2/9 x 5/8	89% 13%	76%
whole numbers to decimals:	work out: 60 ÷ 3 60 ÷ 0.3	80% 28%	52%
	10 x 4 10 x 5.13	90% 58%	32%
	100 x 317 100 x 2.3	58% 36%	22%
	what does the 2 stand for: 521 0.2 0.260 0.412	87% 73% 64% 53%	14% 9% 11%

positive numbers to negative numbers:	Work out:	$+2 + +6$ $+8 + -4$	97% 87%	10%
		$+6 - +8$ $-6 - +3$	70% 36%	34%
	given route on a diagram, express as the vector:	$(2\ \underline{\ }1)$ $(1\ \underline{\ }1)$ $(0\ \underline{\ }3)$	76% 63% 54%	13% 9%
known numbers to unknown numbers:	number of sides $\rightarrow$ number of diagonals from vertex of polygon:	$57 \rightarrow$ $k \rightarrow$	75% 52%	23%
	area of drawn rectangle of dimensions:	6 and 10 5 and $e+2$	89% 12%	77%

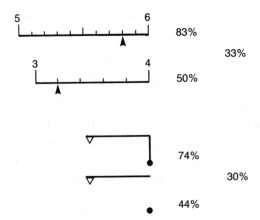

In other cases the difficulty lies not so much in the nature of the elements but in the need to construct some element not immediately given. For example, both the items above involve tenths, but in the first item the tenths are given, whereas in the second it is necessary to 'construct' them rather than simply count the elements given (33 per cent gave the answer 3.1). A similar idea applies to the second of these rotation items, where it is necessary to 'construct' a line from the centre to the flag.

For graphs, difficulty depends strongly on how far the graph corresponds to the visual appearance of the situation represented.

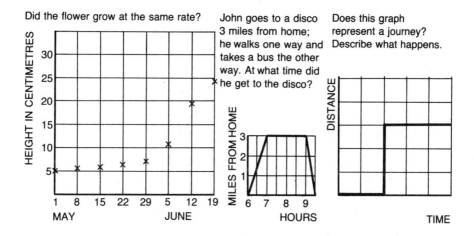

Did the flower grow at the same rate?

John goes to a disco 3 miles from home; he walks one way and takes a bus the other way. At what time did he get to the disco?

Does this graph represent a journey? Describe what happens.

For the first items shown in the above diagram, there is a direct correspondence between the graph and the situation it models: the horizontal axis corresponds to the ground, and the height of the crosses corresponds to the height of the flower. In contrast, the appearance of the other two graphs is misleading: thus in the second graph John does not travel up a hill, along a plateau and down again, although it is fairly easy to establish a correspondence between the three given routes of the journey and the three distinct sections of the graph; in the third graph there is no indication of what the journey might be, so that the reliance on the graph's appearance is likely to be particularly strong (though it should be said the item did not correlate very well with the graphs test as a whole, so there may also be quite different factors causing the massive increase in difficulty).

The *absence* of a diagram may also make an item more difficult, as seems to be the case in the second of these vector items:

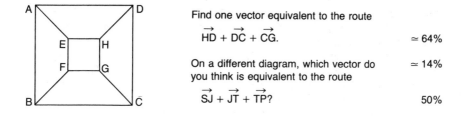

Find one vector equivalent to the route

$$\vec{HD} + \vec{DC} + \vec{CG}.$$ ≈ 64%

On a different diagram, which vector do you think is equivalent to the route ≈ 14%

$$\vec{SJ} + \vec{JT} + \vec{TP}?$$ 50%

Another way of looking at vertical differences is to consider the operations involved. Here one might examine the type of operation, the number of operations, the degree to which they have to be co-ordinated, and whether they are explicit or implicit:

subtraction v multiplication:	write a story for		
	84 - 28	77%	46% (11 yr olds)
	34 x 28	31%	
co-ordination of operations:	Add 4 onto 3n	36%	19%
	Multiply n + 5 by 4	17%	
explicit or implicit:	Ratio : eels x $1\frac{1}{2}$	50%	30%
	K's x $1\frac{1}{2}$	20%	
	5/8 x 2/3	47%	
1 km is the same as 5/8 mile. How long is			39%
	2/3 km in miles	8%	

There is a close relationship between the element dimension
and the operation dimension: sometimes they are simply
alternative ways of describing the same phenomenon: for
example the element 'fraction' can be thought of as a
co-ordination of pairs of whole number; in the case of ratios,
the key elements (namely the multipliers) are operations; at
other times, the nature of the operation determines the way
the elements can be interpreted: e.g., when simplifying
$2a + 5b + a$ the elements can be thought of as names for
objects (apples and bananas, say) but this does not make sense
in $3a - b + a$ ($\uparrow$13 per cent).

It may be the case that the various dimensions discussed in
this section can ultimately be explained in terms of the
single construct 'processing capacity' (see Pascual-Leone,
Case). for example, just as the task of co-ordinating several
operations can be said to impose a greater cognitive load
than working with a single operation, so it may be necessary
when working with unfamiliar or abstract elements to construct
additional, more familiar elements to support them (e.g., a
child working with a specific unknown may have to tell himself
that it is like operating on numbers such as 5, 7, 293, etc.).
However, this reduction to a single construct is not practicable at the present time, and it is likely to be more fruitful
to build up a set of dimensions that can be easily identifiable, even if they sometimes overlap.

Horizontal comparisons are more difficult to make sense of
than vertical ones. For example, to answer the question 'Do
children who understand decimals also understand fractions?'
it is first necessary to define the *kind* of understanding that
is being considered: on the Decimals test, children 'at the
75 per cent level' (i.e., who can in general cope with the
items of facility 75-100 per cent but not the harder items) no
longer ignore the decimal point, and they recognize that 0.2
means two tenths, but they have not yet integrated decimals
into a system, e.g., they may well say that the 2 in 0.250
represents hundredths. Thus the answers to horizontal questions need to be heavily qualified. Also, in examining an item
of a given facility it is difficult to determine which of its
many features are important. However, when the items were

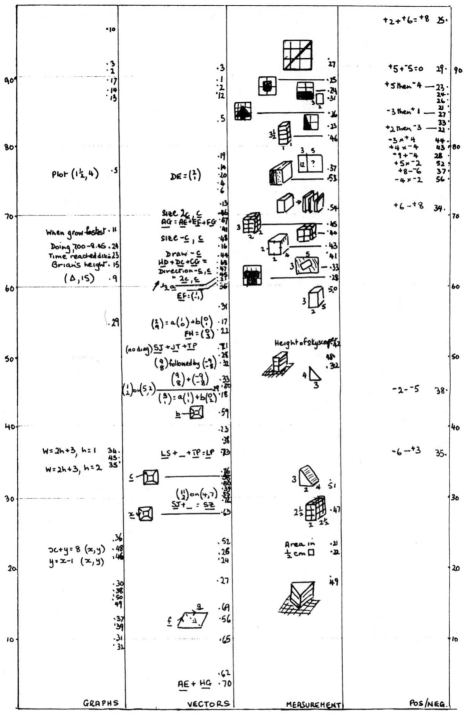

General Sample - Third Year Facilities

63

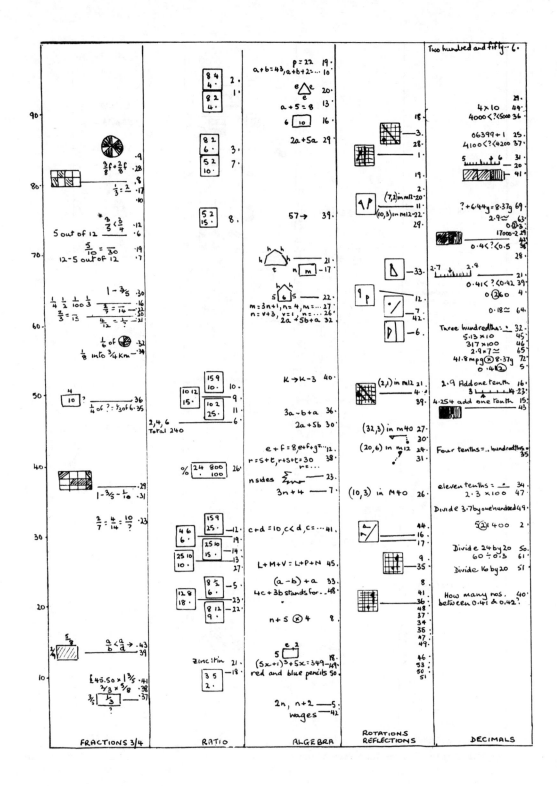

FRACTIONS 3/4   RATIO   ALGEBRA   ROTATIONS REFLECTIONS   DECIMALS

constructed a deliberate attempt was made to minimize extraneous features (such as obscure wording or the need for complex
algorithms) and this has helped to make horizontal comparisons
worthwhile.

An examination of the diagram on pp.63-4 suggests that a
useful way of partitioning the items is to draw three lines at
roughly the 75 per cent, 50 per cent and 25 per cent facility
levels. The discussion that follows will examine the similarity, across tests, of items at each of these levels, as well
as considering the changes in cognitive demand that occur as
one moves from items at one facility level to items at the
next. To simplify the discussion it will be assumed that
children who can answer items of a given facility on one test
can answer items of the same and higher facility on the other
tests.

From the easiest items down to the 75 per cent facility level.
As one moves down one can observe a gradual extension of the
whole number system to less familiar elements which can be
formed by co-ordinating simple whole numbers. Children who
can solve items at the 75 per cent facility level can use
co-ordinates to represent a point, vectors such as (2 1) and
(O 7 4 3) to represent a route or an ordered list, fractions
such as 3/8 (not 3/5) to represent shaded areas, and decimals
to represent subdivision of a line.

In some circumstances such children can also operate on
these elements: for example in one of the eels ratio questions
they recognize that the correspondence 5, 2 (5 cm eel: 2
sprats) is in some sense equivalent to 15, 6; they also know
that 1/3 = 2/6, and they can add 3/8 and 2/8 when the fractions
apply to a realistic context. However, these elements are not
yet integrated into a system: thus, they are unable to
establish the equivalence of less familiar fractions (2/3 =
?/15) or to differentiate tenths from other decimals, or to
form route vectors such as (1 $^{-}$1) or (O $^{-}$3). At this facility
level, children can add integers by co-ordinating shifts
along the number line, but what is perhaps more significant is
that for subtractions like $^{+}$6 - $^{+}$8 they are beginning to move
out of the domain of positive numbers, rather than simply
taking the smaller number for the larger.

In graphs, children
can interpret isolated
features when these cor-
respond directly to
reality: e.g. they
recognize that Frank is
'short and fat' but not
that Helen is of a simi-
lar appearance. Also
children are beginning
to be able to *relate*

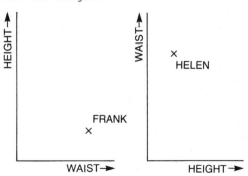

such intuitable features, as in the graph showing the weekly
height of a flower, when asked to identify when it grew
fastest.

In measurement, children
can determine areas, not only
by counting squares of half-
squares, but when other simple
fractional parts are involved,
as in the adjacent triangle,
where some kind of trans-
formation is required. They
can also determine the area
of a rectangle, but not a
triangle, by multiplication,
and they can determine volume by counting cubes when all are
visible.

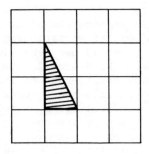

In transformation geometry children have internalized the
action of folding sufficiently to be able to construct images
that involve more than one step, but only when those steps can
be performed in a natural sequence, and when guided by a grid.

From the 75 per cent down to the 50 per cent facility level
As one begins to consider children who can solve items below
the 75 per cent facility level, the understanding of these
various elements is extended. In vectors, for example,
children begin to enlarge the field of route vectors from
those involving positive whole numbers only, to negatives and
zero. However, when asked to combine routes they are likely
to count squares rather than add the vectors, and it is not
until children can cope with items at about the 50 per cent
facility level that a clear move away from directly untuitable
reality can be observed. For example, children can now
describe the effect of a translation (9 8) followed by a
translation of ($\bar{9}$ 8) (the first vector goes beyond the graph
that the children are given), and they can find a vector
equivalent to the route $\overrightarrow{SJ}$ + $\overrightarrow{JT}$ + $\overrightarrow{TP}$ without the aid of a
diagram.

In measurement, children begin to be able to count hidden
cubes and use a formula for the volume of cuboids, but it is
only when the 50 per cent facility level is reached that they
can cope with such figures when the dimensions are not
directly given and it is only then that they seem able to
enlist the aid of an imaginary rectangle in order to work out
the area of a triangle.

In ratio, children at the 50 per cent facility level are
no longer restricted to applying simple operations (repeated
addition or integer multiplication) to the given correspondence (e.g. 5,2 → 10,4 → 15,6) but can make use of new
correspondences when these are prompted by their method, e.g.
the process of building up from 10, 2 to 20, 4, in order to

66

determine 25, ?, reveals the need for the additional
correspondence 5, 1, which these children are able to construct and use.

In generalized arithmetic children begin to accept
unclosed answers like $4h + t$ (rather than $4ht$ or $5ht$) and
$3a + 5b$ (rather than $8ab$) in cases when the letters can be
treated as objects, but it is not until the 50 per cent
facility level is reached that they begin to construct such
expressions for unknown numbers (e.g., a figure with k sides
has $k - 3$ diagonals).

In transformation geometry children begin to judge simple
reflections far more critically,
for example, they recognize that
the lines in the adjacent diagram,
despite their symmetry, will not
reflect one flag onto the other.
However, it is not until about
the 50 per cent facility level
that they can cope with reflections that go off the page, or
with rotations where there is
no link between the object and
the centre.

In decimals, children begin to differentiate tenths from
other decimals and are able to use and impose an ordering on
progressively less intuitable decimal fractions (tenths,
hundredths, thousandths, etc.). However, it is not until
about the 50 per cent facility level that they begin to relate
decimals to each other quantitatively; e.g., they now recognize that one-tenth added to 2.9 is 3.0 and not 2.10; and
that entities smaller than one are not simply recorded in the
first available decimal place (e.g., the answers to the items
in the diagram below are not 3.1 and 1.7).

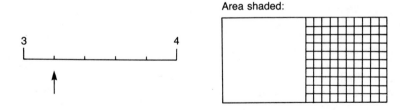

Area shaded:

From the 50 per cent down to the 25 per cent facility level.
As one moves gradually down to about the 25 per cent facility
level, this ability to cope without direct 'concrete' support
is consolidated. Children can determine the missing route
vectors in $\overrightarrow{LS}$ + $\overrightarrow{}$ + $\overrightarrow{TP}$ = $\overrightarrow{LP}$, and then in $\overrightarrow{SJ}$ + $\overrightarrow{}$ = $\overrightarrow{SZ}$, as

67

well as in the more direct S$\dot{\text{J}}$ + JP + TP = $\rightarrow$, without the
aid of a diagram. The understanding of the relationship
between the area of a triangle and a rectangle is extended to
being able to find the volume of a triangular prism. Children
can also go beyond the direct intuition that halving linear
measurements of a figure halves its area. They can subtract
integers as well as add them, even though subtraction is not
easily modelled and may produce a larger answer than the
original integers. They also recognize that division may
produce larger answers when working in decimals. They perform
reflections analytically, rather than in a step-by-step
manner, and in ratio they can construct mediating correspondences when these are *not* prompted by a building-up approach
(e.g. 15, 9 → 5, 3 → 25, 15). In ratio they are also able to
overcome the tendency to resort to the addition strategy in
cases where the essentially multiplicative relationship
between the correspondences is not obvious (e.g. the geometrical enlargement of K and also in the Mr Short and Mr Tall
item, where the symmetry of the given numbers and the small
difference between them reinforces the addition strategy).
However, it can be argued that the understanding achieved at
about the 25 per cent facility level is still only an
extension ('concrete generalization') of the work with
directly intuitable elements, and it is not until children
succeed with items beyond this facility level that they can
work with abstract mathematical systems *per se*. Thus it is
only when one looks at the still harder items that children
seem able to work with numbers that provide no intuitive
support: e.g., on the fractions test, the rectangle of area
1/3 and one side of length 3/5; the ratio item involving an
enlargement of an L-shape, where the multiplier is 5/3 and
there is no mediating correspondence involving whole numbers.

   In generalized arithmetic children can operate on
operations rather than just on unknowns, and they can establish relationships between the *changes* in the values of
unknowns rather than between the unknowns themselves. Also
in items involving objects, they can avoid the temptation of
using letters to represent the objects. In graphs, they can
establish relationships like $x = 2$, or $y = 2$, to represent an
entire set of points, rather than focusing on, or relating,
isolated features of a graph.

In vectors, children have a
flexible understanding of
vector addition and equivalence, for example they can
make sense of the question
'Find one vector equivalent
to $\overrightarrow{AE}$ + $\overrightarrow{HG}$'. Perhaps the
most graphic demonstration
that children have moved
entirely away from what is
tangible or familiar is that
they can now answer questions

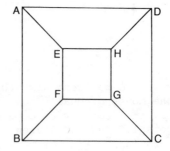

like 'How many numbers are there between 0.41 and 0.42?'.

4. COHERENCE OF THE STAGES

Contextual Variations

There is an increasing number of studies that claim to show,
by changing certain features of classical Piagetian tasks,
that children are able to achieve Piagetian performance
categories (such as 'conservation' or making 'transitive
inferences') at an earlier age than stated by Piaget (see
Gelman 1972, Bryant and Trabasso 1971, Donaldson 1978).
Piaget has generally paid little regard to the effects of such
changes (beyond the acceptance of major decalages, such as
the delay in conservation of weight, say, compared to conservation of number). These studies are useful in as much as
they underline the importance of defining the precise nature
of any task used to assess children's levels of cognitive
development, and help to clarify what is meant by 'succeeding'
on a task. For example, Gelman (1972) shows that the ability
to judge that a row of three elements is greater than a row of
two after the row of three has been transformed (at which some
children as young as three years of age succeed), is very
different from, and very much easier than judging the equality
of *two* rows of three elements in a classical conservation
experiment. Donaldson (1978) discusses a set of tasks using
toys arranged as in the diagram below, in which children are
asked about the number of 'steps' from the Teddy to the chair
and table. The tasks are designed as variants of Piaget's
class inclusion tasks (in which, for example, a bunch of
daisies is compared to a bunch of (the same) daisies and a few
other flowers). The toys task is particularly easy when no
reference is made to the 'steps' and children are simply asked
'Is it further for Teddy to go to the chair, or further to go
to the table?'. Donaldson makes the point that this is no
longer strictly a class inclusion task, but rather one in
which a distance is included in another; however it is a good
example of the sensitivity of these tasks to variation which
one might assume preserved the structure.

Hughes has conducted a very lengthy and thorough investigation of children's understanding of area, weight and volume,
by means of individual interviews, some of which is reported
briefly in Hughes, 1979. The study includes 19 conservation
tasks which were carefully varied in terms of the practical
situations and the materials used, and which were designed to
show 'whether there is an order in the understanding of conservation in weight, area and volume' (Hughes, 1979, p.6).

From his results, Hughes argues that contrary to Piaget's assertion there is not a clear order; instead, 'variables such as expressions used, nature of the task material and method of presentation of the problem ... influence the way the child responds in a dramatic way' (ibid. p.8). However, there would seem to be no obvious reason why, for example, it should not be possible to construct a relatively complex number item, say, which is more difficult than a straightforward volume item. In this brief report, Hughes presents tables showing some of the children's (N = 146, mean age = 7.05 years) performances on selected pairs of items. One such table is given below, which shows that 52 of the 146 children answered both items (Volume 7 and Weight 4a) correctly, whilst 23

Volume 7	23	52	
	52	24	Weight 4a

succeeded only on Volume 7 and 24 only on Weight 4a (the remaining 47 children answered neither item correctly). Hughes feels that the degree of consistency in children's performances is not very great. This is difficult to evaluate objectively, but it is interesting to note that when the correlations (PHI) are calculated for the tabulated pairs, the values (with the exception of item Volume 8) compare favourably with the correlations obtained for some of the 'better' items on the CSMS mathematics tests.

PHI	Wt 4a	Wt 6a	Area 7b	Area 9a
Vol 7	0.36	0.44	0.59	–
Vol 8	0.12	0.24	0.31	0.28
Area 7b	0.36	–	–	–
Area 9a	–	0.51	–	–

It is difficult to accept that the effects of these task-variables contradict Piagetian theory; they show only that the effects of task *structure* on difficulty can be overlaid by the effects of context.

Training Studies

Many of the early studies (cf., Flavell, 1963) that attempted to improve children's performances on classical Piagetian tasks by training met with little success, particularly with respect to retention and transfer; however, there is a growing number of studies, particularly with young children (e.g., Gelman, 1969, Sheppard, 1973, 1974) that claim to have induced improvements in children's cognitive performance which generalize across a range of tasks.

Interestingly, Piaget (1972) has expressed grave doubts about the efficacy of trying to accelerate cognitive growth. However, training (or learning) studies have been carried out in Geneva (Inhelder et al., 1974). These used methods that improved the performance of children who on pre-tests were at the transition to a given stage (concrete operations) but which had little or no effect on children below this stage (pre-operations). These studies are concerned with revealing the *spontaneous* (everyday) learning mechanisms that children employ and as such are not in competition with more direct teaching methods, since 'they are not interested in maximising learning' (Pascual-Leone, 1976). It can therefore be argued that the important issue is not that direct procedures are more successful, but rather to consider the nature of their success and how it comes about.

Pascual-Leone (1976) discusses one such study by Lefebvre and Pinard (1974) who, in contrast to Inhelder et al., use techniques which, in Pascual-Leone's terminology, facilitate the learning of 'executive-schemes' and produce 'the *forced,* i.e., externally induced, co-ordination (chunking) of task-relevant schemes - thus reducing the mental processing complexity ... of the task' (p.275). What this means can best be explained by taking an example, such as a quantity conservation task, say, in which water is poured into different shaped cylinders. One way of mastering this task is by (1) learning to ignore perceptual cues (the effects of changes in height and width are contradictory) and instead to focus on the act of pouring, and then (2) recognizing that this does not change the quantity of water, because 'nothing has been added or taken away' (identity) or because the transformed water can be poured back to its original state (reversibility).

Step 1 (the executive scheme) can be induced by deliberately focusing attention on the attributes of height and width (which makes the *task* easier, since the child does not have to select these attributes for himself); and step 2 can be induced by pouring the water back and forth, by choosing carefully graded examples including beakers of the same size, or by stating the result as a set of rules (which makes the *learning* easier, since the identity/reversibility scheme is reinforced).

Although the changes in children's performance due to such training may be substantial (and possibly worthwhile), the importance of Pascual-Leone's analysis is that it strongly suggests that the effect is *not* a change in the child's cognitive level (or 'M power').

The problem of the degree to which children understand conservation after training, as is indicated by the nature of their responses, occurs in all these studies. Gelman (1969) states that in her study explanations indicating an understanding of reversibility and identity 'occurred frequently'

71

(p.183). Sheppard's (1974) reporting is far more thorough, and he states that '75% of the explanations (after training) emphasise(d) identity' (p.727) and that these 'did not involve simply repeating information acquired in training sessions', so that they represent 'a genuine understanding of the conservation concept'. This study differs from some others in that Sheppard argues that the initial level of operativity *is* of significance in whether or not children benefit from training. The study is also of interest in that Sheppard deliberately tried to train conservation through an understanding of compensation, in which, judging from the quality of children's responses, he did not succeed. At the formal operational level, Lawson and Wollman (1971) trained for the control of variables with bouncing balls and obtained transfer to the pendulum experiment but not to proportional reasoning.

In summary, this section has attempted to make two points.

1. In general it does *not* appear to be the case that training procedures have any substantial effect on children's operational levels (unless children are already at the transition to the stage at which the training is directed). This suggests that taking note of children's levels of cognitive development is likely to be of benefit in the selection and design of teaching materials.

2. On the other hand, there is substantial evidence to suggest that it *is* possible to teach children to solve problems belonging to a higher stage by simplifying the nature of the task and by providing simpler methods of solution. Though this is an obvious truism it does mean that if it is felt desirable to teach children, say, to conserve number or to control variables, it might be possible to do this before children have reached the appropriate level of development, provided it is recognized that appropriate simplifying procedures need to be found. (This is discussed further in Chapter Seven, particularly with reference to Case's work.)

5. STAGES AND THE CURRICULUM

Shayer at Chelsea College has adapted, for class use, five of the Piagetian tasks which provide criteria for distinguishing between concrete and formal operational thinking, and has administered these to large representative samples of children in British secondary schools. He claims that there is a reasonable homogeneity between the different tasks, that is, that they can be said to be measuring some general level of thinking, and that the proportion of British schoolchildren who reach the full formal operational stage before leaving school is not more than about 20 per cent. The graph (page 74), from Shayer and Wylam (1978), shows the proportions of children in English schools who are at the four Piagetian sub-stages: early and late concrete and early and late formal, as measured by three Piagetian tasks (horizontal and vertical, volume and density, the pendulum). It shows a fair degree of correspondence between the proportions of pupils defined as late concrete or early formal on two different tasks, which provides some degree of evidence as to the existence of these stages across tasks for groups of pupils, although not necessarily for individuals. In a later study (Shayer, 1978a), in which about 500 children were tested on five class tasks, the mean correlation was 0.59. Other studies show correlations ranging from 0.3 to 0.8 (Shayer, 1979, Brown and Desforges, 1978). On this basis, Shayer (1979) presents a case for the general usefulness of these stages in considering the demands of the mathematics and science curricula as a whole; but Desforges and Brown (1979) express a contrary view. In this work, Shayer and his colleagues examined curricula and text books in science and attempted to categorize the material according to whether it required formal or concrete thought for its mastery. Further, in an experiment based mainly on the chemistry syllabus, he showed that those pupils who by his measures had not attained formal reasoning did fail on those items on the syllabus which he had identified as requiring this ability (Shayer, 1978b). This work continues.

Attempts to do something similar but less ambitious with the mathematics curriculum were made by Malpas and Brown (1974). They distinguished between concrete and formal models, the former being drawings, photographs, sets of plans (as for a house), a model house, verbal descriptions of objects or events. Formal models are those which represent the relationships between the relationships expressed by individual concrete models, for example, graphs representing functions, formulae representing functions, and so on. There are also intermediate kinds of models, such as bar charts, in which formal relationships are expressed, but in a way in which the physical attributes of the model lend support to the data. Malpas (1974) examined the SMP Mathematics Series A - H and identified the incidence of concrete and formal models in each chapter of the course. He concluded that the demand for formal thinking was high from Book D onwards and that this would be likely to be beyond the level of thinking of all but about the top 20 per cent by ability, or fewer, of the secondary population. The criteria for stage allocation of types of material in this work are broad and imprecise, though they have some plausibility.

A more developed approach to this task is represented by the analysis of the CSMS test results, given earlier in this chapter.

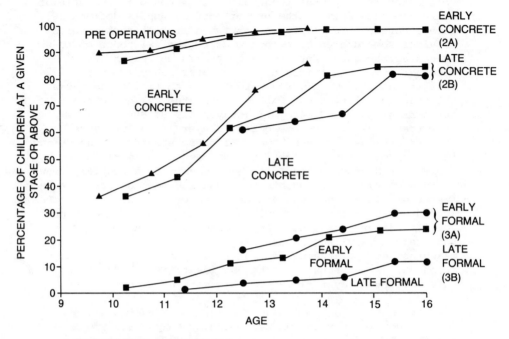

▲ By Task I (Horizontal and Vertical)
■ By Task II (Volume and Density)
● By Task III (Pendulum)

References for Chapter 3

BOYLE, D.G. (1969): *A Student's Guide to Piaget*. Pergamon.

BROWN, G. and DESFORGES, C. (1977): 'Piagetian psychology and education: time for revision', *B J Ed Psych,* 47, 7-17.

BRYANT, P.E. and TRABASSO, T. (1971): 'Transitive inferences and memory in young children', *Nature,* 232, 456-8.

COLLIS, K.F. (1975): *A Study of Concrete and Formal Operations in School Mathematics: A Piagetian Viewpoint*. Melbourne:ACER.

DESFORGES, C. and BROWN, G. (1979): 'The educational utility of Piaget: a reply to Shayer', *B J Ed Psych,* 49, 277-81.

DONALDSON, M. (1978): *Children's Minds*. London: Croom Helm.

FLAVELL, J.H. (1963): *The Developmental Psychology of Jean Piaget*. New York: Van Nostrand.

GELMAN, R. (1969): 'Conservation acquisition: a problem of learning to attend to relevant attributes', *J Exp Child Psych,* 7, 167-87.

GELMAN, R. (1972): 'Logical capacity of very young children: number invariance rules', *Child Dev,* 43, 75-90.

HALFORD, G.S. (1978): 'An approach to the definition of cognitive developmental stages in school mathematics', *B J Ed Psych,* 48, 298-314.

HART, K.M. (1980): *A Report of the Mathematics Component of the CSMS Programme*. London: Chelsea College.

HUGHES, E.R. (1979): *Should We Check Children?* University of Leeds: Cognitive Development Research Seminar, September 1979.

INHELDER, B. and PIAGET, J. (1958): *The Growth of Logical Thinking*. London: Routledge and Kegan Paul.

INHELDER, B., SINCLAIR, H. and BOVET, M. (1974): *Learning and the Development of Cognition*. London: Routledge and Kegan Paul.

LAWSON, A.W. and WOLLMAN, W.T. (1971): *Encouraging the Transition from Concrete to Formal Cognitive Functioning - an Experiment*. AESOP, Lawrence Hall of Science, Berkeley, California.

LEFEBVRE, M. and PINARD, A. (1974): 'Influence du niveau initial de sensibilité au conflit sur l'apprentissage de la conservation des quantités par une méthode de conflit cognitif', *Canadian J Beh Sci,* 61, 398-413.

LUNZER, E.A. (1973): 'Formal Reasoning: a re-appraisal', Keynote Paper. Symposium, Jean Piaget Society, Philadelphia, Spring, 1973. Published in (amongst others) FLOYD, A. (Ed), 1979: *Cognitive Development in the School Years*. London: Croom Helm.

MALPAS, A.J. (1974): 'Objectives and cognitive demands of the School Mathematics Project's main school course', *Maths in School,* 3(5), 2-5, and 3(6), 20-1.

MALPAS, A.J. and BROWN, M. (1974): 'Cognitive demand and difficulty of GCE O-level mathematics pre-test items', *B J Ed Psych,* 44, 155-62.

PASCUAL-LEONE, J. (1976): 'On learning and development, Piagetian style: 1. a reply to Lefebvre-Pinard', *Canadian Psych Rev,* 17, 270-88.

PIAGET, J. and INHELDER, B. (1969): *The Psychology of the Child*. London: Routledge and Kegan Paul.

PIAGET, J. (1972): 'Intellectual evolution from adolescence to adulthood', *Human Dev,* 15, 1-12.

ROMBERG, T.A. and COLLIS, K.F.: The Assessment of Children's M Space. Technical report, University of Wisconsin Graduate School R and D Center, Madison, Wisconsin.

SHAYER, M. (1978(a)): A test of the validity of Piaget's construct of formal operational thinking. Ph.D. thesis, University of London.

SHAYER, M. (1978(b)): 'The analysis of science curricula for Piagetian level of demand', *Studies in Science Ed,* 5,115-30.

SHAYER, M. and WYLAM, H. (1978): 'The distribution of Piagetian stages of thinking in British middle and secondary school children II: 14-16 year olds and sex differentials', *B J Ed Psych,* 48, 62-70.

SHAYER, M. (1979): 'Has Piaget's construct of formal operational thinking any utility?' *B J Ed Psych,* 49, 265-76.

SHEPPARD, J.L. (1973): 'Conservation of part and whole in the acquisition of class inclusion', *Child Dev,* 44, 380-3.

SHEPPARD, J.L. (1974): 'Compensation and combinatorial systems in the acquisition and generalisation of conservation', *Child Dev,* 45, 717-30.

Chapter Four
Skills, Concepts and Strategies

1. The Components of Mathematical Competence

2. Arithmetic Skills and Concepts

   Computation and Recognition of Operation

   Factorial Independence of Computational Ability

1. THE COMPONENTS OF MATHEMATICAL COMPETENCE

In discussing the understanding and teaching of mathematics we must distinguish between skills, concepts, strategies and attitudes. All workers use such terms, but although they communicate at a crude level they do not capture very well the real distinctions between types of material.

Gagné and Briggs (1974) distinguish five components, intellectual skills, cognitive strategies, verbal information, motor skills and attitudes, and discuss for each the conditions necessary to facilitate their learning. A more restrictive scheme comprising computation, comprehension, application and analysis was used as a basis for constructing mathematics tests for the (American) *National Longitudinal Study of Mathematical Abilities* (Begle, 1979). The APU (1980) uses concepts, skills, applications and problem-solving. Davis and McKnight (1979) list procedures and sub/super-procedures, 'visually moderated sequences', frames and heuristics among a total list of ten 'ingredients' - these last three correspond roughly to skills, concepts and strategies, but Davis is here following the present tendency of cognitive psychologists to model mental structures and abilities from an information processing point of view. In offering our own list of categories, we wish particularly to draw attention to the key properties which distinguish them. These are: 1) the kind and degree of connectedness which they possess, 2) whether they refer to observable actions or to inferred contents of memory, and 3) whether they are based on symbols or on the concepts denoted by the symbols.

The term *fact* will be used for items of information which are essentially unconnected or arbitrary, so that they cannot be directly supported by conceptual structure. These include notational conventions, such as the index notation and the fact that 34 means three tens plus four and not four tens plus three; conversion factors, such as that 2.54 centimetres equals an inch; and the names attached to particular concepts, for example that the ratio 'opposite/hypotenuse' is called sine and not tangent. These are hard to remember and sometimes an artificial link for them is provided by a mnemonic, such as SOHCAHTOA for the trigonometric ratios. (The so-called 'number facts' do not fall into this category since they form part of a rich network of connections.) Because we remember connected material very much better than arbitrary facts, mnemonics offer a very proper way of assisting memorization where there are no intrinsic connections in the material which can be used (see the section on Aids to Recall, Chapter Two).

The term *skills* cannot be restricted to the standard computational procedures of arithmetic and algebra. We shall use it to include any well-established multi-step procedure, whether it involves symbolic expressions, or geometrical figures, or neither. For example, the ability to multiply a given number by 10 by moving the decimal point is a very short symbol-based skill, and the ability to multiply 0.03 by 0.2 by separately counting the number of decimal places and forming the multiplication 3 x 2 is a slightly more complex but still short procedure. The finding of the centre of a circle passing through three given points is a multi-step sequence based on diagrammatic features. We call these 'symbol' - or 'diagram' - based because while they are being performed the focus of attention is on the movements or transformations of the symbols themselves, and one is not usually reminding oneself, for example, that the different decimal places refer to tenths, hundredths and thousandths, or that the perpendicular bisector drawn is in fact the locus of all points equidistant from the two points to which it relates. As Davis puts it, these procedures are 'visually moderated', in that the performance of each step produces an expression or figure which cues the next step. Examples of non-visually-moderated skills are harder to find, but an example might be a developed routine for deciding which would have the stronger orange taste out of two mixtures, one made from two glasses of orange to three of water, and another from three glasses of orange to five of water. Thus the essential features of skills are that 1) they are actions or transformations rather than items of knowledge (though they do have mental programs which govern them) and that 2) their type of connection is that of a chain, although sometimes a branching one.

The term *concept* has two meanings in common use - one more precise but less important, the other vague but more important. The first meaning distinguishes a *concept* from a *relationship;* here a concept is a 'thing' (e.g., rhombus, group, multiplication) needing, in formal mathematics a definition, while a relationship is expressed by a statement

(e.g., 'multiplication of real numbers is commutative').
Concepts may be defined extensively, by giving members and
non-members of the class which constitutes the concept, or
intensively, by stating the defining properties or conditions.
The second meaning of the term is that implied when one is
speaking of 'teaching for concepts' or of 'having the concept
of place value'. These are loose expressions - the concept of
place value includes awareness of bases other than ten and of
decimals, but 'having the concept' might simply be intended to
mean being able to handle hundreds, tens and units sums with
understanding. The better term here is *conceptual
structure*, meaning a network of concepts and relationships.
Such structures underpin performance of skills, and their
presence is shown by the ability to remedy a memory failure
or to adapt the procedure to a new situation. The real
importance of the conceptual structure is that as a richly
inter-connected network it constitutes a stable memory
structure, in which any particular link which fades is
relatively easily reinstated. The learning of a new concept
or relationship implies the addition of a node or link to the
existing cognitive structure, thus making the whole, if
anything, more stable than before, whereas the learning of a
skill requires the establishing of a set of more or less
arbitrary links between the steps.

General strategies are procedures which guide a person's
choice of what skills to use or what knowledge to draw upon at
each stage while in the course of problem-solving or investigation, or indeed of verification of a discovery. They, too,
have their associated concepts, such as data and conclusion,
special case, converse, generalization and so on. Thus they
are of the same nature as mathematical skills and concepts
but they act, not upon the data of the given situation but on
the person's own activity. They are 'superprocedures' or
programmes which operate on other programmes. The teaching
and learning of these are considered in Chapter Eight.

Appreciation involves the awareness of the nature of
mathematics and the affective response to the subject.
Children may perceive mathematics as a set of techniques or a
means of finding patterns and solving problems, as somehow
being relevant to social needs or as underlying a wide range
of practical situations. Included under this heading are
'attitudes' towards mathematics, the liking for the subject,
the confidence in it and the esteem in which it is held.
Chapter Nine deals with these.

The distinctions we are drawing here relate to different
types of mathematical knowledge and skill as possessed by an
individual. However, which type of capability will be brought
into play for a given task depends on the individual as well
as on the task. For example, Brown (in Floyd, 1978) considers
the task:

'Express 3/8 as a decimal'.

This may be solved by:

(a) *factual recall,* provided the knowledge that 3/8 = 0.375 is readily accessible in the individual's memory;

(b) a *skill,* performed according to a known, standard computational procedure for converting fractions to decimals;

(c) the use of a *conceptual structure* relating fractions to decimals in some way. It may be known, for example, that 1/4 is 0.25, so 1/8 is half of this, i.e. 0.125, so 3/8 is 3 x 0.125 = 0.375;

(d) a problem, requiring the development of a *strategy* especially in a case where there is some knowledge of the properties of decimals and fractions separately, but no clear initial conception of how to relate them.

It is this fact which renders hazardous, if not nonsensical, the attempt to categorize test items in themselves under such headings.

2. ARITHMETIC SKILLS AND CONCEPTS

Computation and recognition of operation

A pair of skills and concepts which are confused more often than they should be are those needed to solve the common 'verbal problems' of arithmetic, in which it is necessary to decide what numerical operation is required and then to carry out the appropriate computation.

Only comparatively recently has it been recognised that the recognition of the correct operation in these cases can be a matter of significant difficulty. Rees (1973) has analysed the difficulties in mathematics experienced by craft and technician students. Even students who are competent in computational skills often find it difficult to identify the appropriate calculation in verbal craft items: many are unable to divorce the calculation from the context so as to write it down in a tractable form. As one student commented, 'It (i.e., the craft context) makes the stuff more interesting, but it's difficult to sort out the calculation from all the junk'.

The last comment leads into the discussion of the nature and purpose of word problems. The part played by such problems in the teaching of mathematics is frequently not made clear: it is often suggested that they are intended as means of making the arithmetic more interesting or plausible, a way of dressing up the problems. They are seen somehow as part of knowing how to do the calculation; but they are quite definitely a different kind of exercise from numerical

computations, and require another kind of competence. For
example, two of the tests produced by CSMS Project at Chelsea
College include a number of word problems in which the appropriate calculation is to be chosen (though not performed): one
example from these tests asked for the cost of a packet containing 0.58 kg of minced beef when minced beef is priced at
88.2 pence for each kilogram. The correct calculation '0.58
x 88.2' was chosen by less than 30 per cent. No actual computation is required here, so that this in itself shows that
recognition of an appropriate operation is not something to
be taken for granted: certainly it cannot be used merely to
add interest or plausibility to a numerical calculation. In
another of the Chelsea tests, concerned with the understanding
of operations on natural numbers by children aged 10-13, some
results which compare this ability with pure computational
ability are available. Tests were made in the first year of
the comprehensive school, comparing ability to select the
right calculation with ability to perform the calculation.
Correlations between the results of these tests are fairly
low (0.4 for $\pm$ and x, and 0.27 for $\div$). These results seem
to indicate fairly conclusively, if it was not already clear,
that being able to multiply and knowing when and what to
multiply are not the same kind of ability and are not closely
related to one another, and that both need teaching. About
1500 children aged 10-13 were given five 'sums': 84 - 28,
9 x 3, 84 x 28, 9 $\div$ 3, 84 $\div$ 28 and asked to write a 'story'
to match each expression. Teachers had previously discussed
stories for a trial item (9 + 3), so that the children knew
what was required. A traditional computation test was also
given to some of the children, and here the scores on the
test as a whole and the story test correlated quite highly
(between 0.6 and 0.7). However, detailed study of individual
children's performance in the different tests indicates that
while some children master the algorithms with only partial
understanding of the operations, those with hardly any
grasp at all of the nature of the operations are unsuccessful
with the calculations. On the other hand, there are some
children with a very good understanding of the operation
whose computation is not as accurate as it might be (CSMS,
1980).

A sample assessment of mathematics in secondary schools
in the United States conducted by the National Assessment of
Educational Progress (NAEP) shows some differences in attainment between computational skill and conceptual understanding.
The survey shows that 13-year-olds can do about as well as
adults on most computational tasks and 17-year-olds can do
better.

'Weathermen estimate that the amount of water in nine
inches of snow is the same as one inch of rainfall. A
certain Arctic island has an annual snowfall of 1,602
inches. Its annual snowfall is the same as an annual rainfall of how many inches?'

81

The correct answer was found by 31 per cent of the 13-year-olds, 53 per cent of the 17-year-olds and 58 per cent of the adults: the pattern of attainment here is quite different from that for pure computational ability. On the other hand, the problem is rather complex in terms of the information load and the unfamiliarity of context, and the improvement of scores with age may be a reflection of increasing ability to handle this complexity.

An analysis of attainment is also provided by the APU Primary Survey already mentioned: it is difficult to draw any conclusions about relative attainment in skills and conceptual understanding because of the lack of comparable questions or of detailed description of the questions. However, in division of whole numbers, the question $84 \div 4$ had a facility of 70 per cent, while word problems involving a calculation of roughly comparable difficulty (i.e., not merely a recall of table facts) had facilities varying from 25 per cent to 66 per cent. This again indicates that the difficulty involved in the word problem is of a different kind, and varies according to the context, a feature to be further discussed in the next chapter.

Factorial independence of computational ability

Factorial analyses of mathematical attainment also generally show a separate factor for numerical or computational skill. In Wrigley's (1958) investigation, a factor (called 'numerical ability') was identified upon which depended performance in arithmetic (especially mechanical arithmetic), and to a lesser extent performance in algebra (which in this case consisted of standard manipulative techniques). (It should however be noted that (general) 'high intelligence' is quoted as the most important single factor for success in mathematics.)

Evans (1968), in another factorial study, in addition to factors depending on a particular teacher and on general reasoning ability of the pupil, identified two further factors, one for problem-solving and one for numerical skills. These studies provide some empirical evidence that at least some of the categories identified above are learnt by different pupils with different degrees of success (see Peel in Servais and Varga, 1971).

Begle (1979) states that 'a number of studies have been carried out using a variety of subject-matters that together demonstrate that the six levels of the Bloom taxonomy are empirically as well as conceptually distinct.' He does not give references for these. However, he does report one of the parts of the National Longitudinal Study of Mathematical Abilities (Report No.27), which by an extensive use of regression procedures showed that students' improvements in computation and in the higher-level acquisitions

82

(comprehension, application and analysis) developed relatively independently of each other with not very much interaction. He concludes, 'we thus have empirical evidence that computational achievement is something quite different from achievement at higher cognitive levels.'

Not all the factorial studies yield the same set of components, but at least the relative independence of computational achievement is generally established.

References

APU (1980): *Mathematical Development: Primary Survey Report No. 1.* HMSO.

BEGLE, E.G. (1979): *Critical Variables in Mathematical Education.* Washington: MAA/NCTM.

CSMS Mathematics Team (1980): *Children's Understanding of Mathematics (11-16).* London: Murray.

DAVIS, R.B. and McKNIGHT, C. (1979): *The Conceptualisation of Mathematics Learning as a Foundation of Improved Measurement,* Development Report No. 4, Curriculum Laboratory, College of Education, University of Illinois.

EVANS, G.T. (1968): 'Patterns and development of the mathematical performance of secondary school students', *Aust J Ed,* 153-166.

FLOYD, A. (1978): *Cognitive Development in the School Years.* An Open University Reader, Croom Helm.

GAGNÉ, R.M. and BRIGGS, L.J. (1974): *Principles of Instructional Design.* New York: Holt, Rinehart and Winston.

NAEP (1975): Results and implications of the NAEP mathematics assessment: secondary school. *Math Teacher,* October.

REES, R. (1973): *Mathematics in Further Education: Difficulties Experienced by Craft and Technician Students.* Brunel F.E. Monographs, 5, London: Hutchinson.

SERVAIS, W. and VARGA, T. (1971): *Teaching School Mathematics.* Harmondsworth: Penguin Books.

WRIGLEY, J. (1958): 'The factorial nature of ability in elementary mathematics', *B J Ed Psych,* 28, 61-78.

Chapter Five
The Teaching of Facts and Skills

1.   INTRODUCTION

The key to decisions about what teaching methods to adopt for facts and skills lies in their properties as described above (Chapter Four) that is, primarily on the kind of connectedness which it is desired to establish. In fact, this chapter might be better entitled 'Teaching for Recall and Chaining' and the later one 'Teaching for Interconnectedness'. For immediacy of recall of facts and fluency of chaining of skills, drill and practice are effective, though their effects are short-lived. The value of providing connecting matter in the shape of mnemonic aids for intrinsically disconnected facts has been mentioned already (see Chapter Two, section on Aids to Recall). However, for retention and transfer, the establishment of all possible connections in the memory structure is desirable. Such support for arithmetical and algebraic skills is provided if symbolic processes are related closely to the corresponding conceptual transformations, that is, if the meaning of the symbol-manipulation in terms of what is happening to the denoted concepts is understood. This is the explanation of the superiority of *understood* over *rote-learnt* skills. These points come out clearly in the research of Brownell and others, discussed below (see also Resnick, 1980).

2.   DRILL AND MEANING IN ARITHMETIC

It is not inappropriate at the present time to open a discussion of teaching methods by looking back and considering the extensive research done in the period 1900 - 1950 on the learning of arithmetic. An important paper by Brownell, a leader in the campaign in the United States for meaningful, rather than rote, learning, gives a fascinating sketch of the scene in 1935. He is discussing the place of 'drill' in the basic number facts:

> Research on the effects of drill is now by no means as popular as it was at one time. Experimental interest in the problem began with Thorndike's 1908 study, reached its peak in the period 1910 to 1920, and has now considerably subsided. In more recent years the large number of investigations on drill have been less concerned with its effects upon learning than with such related matters as the length of the drill period, the comparative merits of mixed and isolated drill organisation, and the like. It is as if the instructional value of drill has been firmly established and now the major problem is how the drill should be administered.

> There would seem to be in the experimental literature abundant justification for belief in the vital importance of drill in arithmetic instruction. Without exception, investigations have reported that drill, of whatever sort and however administered, improves efficiency. (Ashlock and Herman, p.170, Brownell and Chazal, 1935.)

But, Brownell asserts, the investigators had not considered the place of drill in the total programme of teaching arithmetic; moreover, 'the implications of their findings have been by no means restricted to circumstances which correspond with their investigations.' In fact, most of the 16 investigations available to Brownell had been made with pupils aged ten or over, yet the teaching of number facts by drill was common from the age of six upwards. It is the uselessness of this *premature* drill, before and without understanding of the meaning of the operations, which Brownell was exposing in this paper. The 63 children in his experiment were tested ten days after entering Grade 3 (the 8+ class) on all 100 addition facts. They had been drilled in all these and the corresponding subtraction facts throughout their two previous school years, yet their median number of errors was still 11 out of the 100. After a month of drill, at five minutes every day, and a further month *without drill*, both speed and accuracy improved. But Brownell was also interested in the extent to which they were acquiring the understanding of relationships, such as '9 + 8 = two 8s (16), + 1, 17'.

He interviewed the children at each stage and classified them as operating in one of four ways: Guessing (incorrect and not subsequently corrected), Immediate Recall (correct), Counting, and Indirect Solution (using relationships). About 45 per cent of the pupils used immediate recall, about 20 per cent guessed, 20 per cent counted and 15 per cent used indirect solutions, and *these proportions remained substantially constant over the three interviews*. More details and

some transcripts appear in the article. Brownell concluded
that

> In spite of long-continued drill children tend to maintain
> the use of whatever procedures they have found to satisfy
> their number needs;
>
> ... drill is exceedingly valuable for increasing, fixing,
> maintaining and rehabilitating efficiency otherwise
> developed,

but
> drill makes little, if any, contribution to growth in
> quantative thinking by supplying maturer ways of dealing
> with numbers ...   (p.186).

In a later large-scale experiment involving some 36
eight-year-old classes, in three different areas, Brownell and
Moser (1949) compared the teaching of subtraction by meaningful and rote methods. Both 'equal additions' and 'decomposition' were taught. The meaningful teaching of the equal
additions method began with the discussion of the principle,
illustrated by such examples as 4 - 1 = 3, 14 - 11 = 3,
24 - 21 = 3, etc., set up with objects and number drawings.
Then, when writing the sums, the carried figure was always
written in, thus:

$$3 \quad 4 \quad {}^{1}6$$
$$2 \quad {}^{4}\cancel{3} \quad 9$$
$$\overline{}$$
$$1 \quad 0 \quad 7$$

The meaningful decomposition teaching was similar, the numbers
being decomposed and the new figures written in, as they
generally are by pupils today. The mechanical or rote teaching gave no explanations, but simply developed a verbal
pattern such as: 'I can't take 9 from 6, so I think of it as
16; 9 from 16 is 7. Since I thought of 6 as 16, I must think
of 3 as 4 ... '. The new figures were not written in. The
teaching period lasted 15 days, after which the children were
given an immediate post-test, including some items similar to
those taught (these involved borrowing only from the tens
column), and some transfer items which required borrowing
either from the hundreds column, or from both tens and
hundreds. Six weeks later they took a retention test.
Measures of both accuracy and speed were obtained in these
tests. Interviews were also conducted at these two times, to
evaluate the degree of understanding of the processes. The
greatest differences between the groups were on the transfer
items, where the group to whom the decomposition method had
been taught meaningfully achieved a mean score of 7.5 items
correct out of 15, compared with 3.1, 2.8 and 2.6 items for
the other groups. The next greatest difference was in degree
of understanding, where the two meaningfully taught groups

scored significantly better (at the 5 per cent level) than the others. Differences in accuracy and speed were smaller, but the rote-taught decomposition group performed somewhat less well than the other three for accuracy, and the rote-taught equal additions group were slightly faster than the others. (These results relate to one of the three centres, but they are consistent with the results of the other two.) Brownell concluded that the decomposition method should be taught, meaningfully, but that the desirability of changing over to equal addition at the age of about ten should be researched, in view of the greater speed attained with this method. It had been clear that the rationale for this method could not generally be understood by the eight-year-olds, but it seemed possible that it might be understandable by ten-year-olds. It might be thought that such a switch would cause confusion. However, Cosgrove (1957), working with 12-year-olds of lower ability, found them able to make this change without lasting interference effects (quoted by Williams, 1971). Nowadays, of course, the question would be complicated by considerations of possible calculator use for larger calculations.

Another thorough study, by Brownell, with Carper (1943), of the learning of the multiplication facts indicates some of the ways in which this body of knowledge can and should be made meaningful before being consolidated by drill. Williams (1971) summarizes their conclusions as follows:

'The verbal form two 3's = 6 is preferable to two times 3 = 6. The tables certainly should be learnt, and in this way: combinations should be presented first of all in random order, and then children should order them; drill might be used for increasing the immediacy of recall of combinations, but should always follow adequate meaningful learning; applications of the commutative law (for example, 3 x 4 = 4 x 3) should be made clear; the properties of 1 and 0 should be taught; the concept of even and odd numbers should be used, especially for showing that even multipliers always produce even products; the special properties of the products of 5 and 9 should be pointed out; combinations should be learnt in pairs (excepting in the case of doubles).' (p.52.)

To summarize: the effectiveness of drill, with feedback, for improving speed and accuracy in the short term is clearly substantiated; so is its ineffectiveness for developing longer-term understanding or more meaningful techniques. A survey by Williams (1971) of later research on this question indicates that the bulk of research favours meaningful learning. It also shows some differential effects, for example, meaningful learning is more helpful for retention and transfer to new tasks, especially more difficult ones, and affects attitudes favourably. Other work showed that meaningful methods followed by drill produced better results than either method on its own. Williams concludes:

> 'When the primary school mathematics teacher is considering
> the controversy over meaningful versus rote methods of
> learning, he is quite likely to be considering it in the
> light of the practices that he would adopt in teaching the
> multiplication tables. Probably most teachers would agree
> that it is better for the child to understand whatever he
> can, but many would claim that the abolition of rote
> learning would lead to inefficiency in the recall of
> multiplication facts. An experienced teacher's views on
> the learning of multiplication tables might go somewhat
> thus: while the child should appreciate the manner in which
> the tables are built up, the facts contained therein should
> be instantly available to him and it should be possible for
> him to bring them to mind without thinking. Unless these
> facts *are* readily available, attention must be given to
> working them out while 'on the job' and this attention
> might be better employed in the mathematical activity for
> which these facts are needed.'

3. COMPUTER-ASSISTED SKILL LEARNING

Evidence (e.g., Nitsch, 1977) suggests that same-context
types of practice increase retention but do not aid transfer,
while varied-context examples aid transfer but are liable to
cause confusion in the initial stages of learning. Thus a
hybrid practice menu is likely to facilitate the ease and
speed of initial learning and also aid flexible transfer on
subsequent tasks.

A well-documented principle which is important for the
management of practice is the influence of feedback. However,
in the human learning situation, the prime importance of feedback is not as a mode of reinforcement, but as a way of
providing information to enable the student to correct errors
(which explains why Grundin, 1962, in a review of 13 studies
found not a single one that showed significant differences in
favour of feedback). Rather than designing small-step
practice programs which guarantee an extremely high level of
success, the task difficulty should be matched to student
competence so that his misconceptions are revealed. Then
appropriate feedback and supporting practice can be given
(see Anderson *et al.*, 1971, 1972).

Two experiments relating to the conditions of practice
of arithmetic skills have been conducted by Hartley and
colleagues at Leeds. In the first of these a computer was
used to provide pupils of seven to eight years with exercises
in addition and subtraction at an individually specified level
of difficulty. This was ascertained from the pupil's previous
results. Three groups of pupils were used and the computer
program was arranged so as to generate examples which maintained a working level of success, for the pupils of one group,

of 90 per cent, for a second group of 75 per cent and for a third of 60 per cent. The pupils worked for 12 sessions of about 12 minutes each, and pre- and post-tests were given. The results clearly favoured the intermediate difficulty group (75 per cent) followed by those with the 90 per cent level. The treatment with the 60 per cent success rate was easily the worst and was not liked by the pupils (Woods and Hartley, 1971). The second experiment investigated different forms of feedback message to pupils following mistakes in arithmetic tasks, this time on multiplication of up to three-digit numbers by a single-digit number.

In another study (Tait et al., 1973) the type of feedback was varied between 'passive' and 'active'. The passive type simply gave a feedback message, then continued with the next question. The active type took the pupil through the question, requiring him to try each step again until he succeeded. The active was more effective for the less able pupils but for the more able the two types were more or less equivalent (but significantly better than no feedback).

In designing feedback programmes for a given task, it is clearly important to identify the kinds of difficulties children experience. O'Shea constructed a programme which was designed to make this identification itself, as well as to provide appropriate feedback. The programme had some success when used with schoolchildren; it also became 'experienced' and improved its teaching performance. (See Hartley, 1980.)

4. SYMBOLS AND MEANINGS

Structurally equivalent symbolic and conceptual tasks are not necessarily recognized as the same. It is desirable that they should be, so that connections are established which help one skill to support the other. The following research shows the relatively low correlation between such a pair of tasks among pupils who are currently learning them.

Steffe and Parr (1968) tested groups of children on various problems involving fractions and ratios. An under-standing of these concepts was tested in terms of the recog-nition, representation and manipulation of various pictorial situations, and then tests were given in a symbolic notation in which operations had to be performed by manipulating the symbols.

Thus in the first test the following items appear:

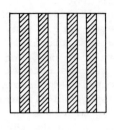

Look at the square in the top picture. Four of the ten equal parts are shaded. Now look at the bottom picture. This square must have the same amount shaded.

How many of the five equal parts should be shaded so that the same amount will be shaded in both squares? The squares are unit squares.

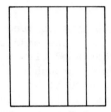

and:

If there are six triangles for every fifteen circles, how many triangles would there be for five circles?

The later, corresponding items are:

$$\frac{4}{10} = \frac{\square}{5} \quad \text{and} \quad \frac{6}{15} = \frac{\square}{5}$$

Find the number that goes in the box.

It seems likely that at least by some pupils the symbolic questions were treated as requests to manipulate the fraction symbol according to the rule 'divide top and bottom by the same number'. It is also possible that in the diagrammatic case of the fraction item the answer was obtained by filling up the total area in the second square equal to that shaded in the first. If this is so, in neither case was the concept of a fraction necessarily involved in the thinking. The low correlation (about 0.37) between pupils' results on the two forms of test confirm that to the pupils the two types of problem are substantially different.

In the CSMS investigations on the understanding of fractions, another situation is provided in which understanding in a conceptual context can be compared with learning a computational technique. Each test paper presents problems and in addition a set of computations (designed to mirror the problems): thus a problem might be

'Shade in 1/6 of the *dotted* section of the disc. What fraction of the whole disc have you shaded?'

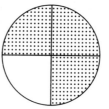

The comparable computation simply reads:-

'1/6 of 3/4 = ...

Tests were given to children in the first four years of secondary school, and in each case the computations were given immediately after the set of problems had been finished. In many cases, the problems proved easier than their comparable computations, pointing to the fact that children used strategies other than the algorithms they had been taught. No connection existed between the problem and the 'sum' in the minds of many children who could successfully deal with the problem but did not try to apply the same method to the computation. It was as if two completely different types of mathematics were involved, one where the children could use common sense, the other where they had to remember a rule.

In addition and subtraction computations, the percentage of first-year pupils succeeding was always higher than in any other year. The ability to carry out these fades as the child gets older, whereas the ability to solve the problems does not decrease with age, emphasising that this is a different and more fully assimilated acquisition, compared with the computational algorithms. In general, children do not rewrite problems in terms of a 'sum', but produce methods which appear to fit the situation.

5. STANDARD ALGORITHMS AND CHILDREN'S METHODS

In recent years some disturbing observations have shown children holding official mathematics and commonsense calculation in separate compartments. Other observations, growing in volume, show how children, though taught standard processes, discard these, partly or totally, in favour of their own *ad hoc* methods based on their understanding of the number system.

In a discussion entitled 'Arithmetic as an arbitrary game', Ginsburg (1977) quotes many examples in which children acquire skill in procedures for performing computation quite separately from any understanding of the notation. Thus, a nine-year-old was asked to add 123 + 52 + 4 and wrote:

$$\begin{array}{r} 123 \\ 52 \\ 4 \\ \hline 219 \end{array}$$

She was corrected, and then performed the calculation properly, but when asked which answer was best she still preferred 219. To her the choice was arbitrary, a matter of caprice. Another girl, given the same problem lined up improperly, was prepared to do it as given without query:

$$
\begin{array}{r}
123 \\
52 \\
4 \\
\hline
683
\end{array}
$$

She said, 'If you do it this way, it's a different number and if you do it another way it's a different number.' A particularly striking example is provided by one pupil who easily and correctly computed 123 x 5 on paper, but was unable to multiply 12 x 6. She saw the first problem as an application of a familiar technique, but attached no meaning to either.

Plunkett (1979) quotes several examples of calculations performed with understanding in a non-standard way. The calculation 'What's 213 take away 188?' is solved: 'Well it's 12 up to 200 and 13 's 25'. Jones (1975) made a more substantial analysis of the methods by which the subtraction 83 - 26 was correctly evaluated. Three of these methods had been taught as standard and twenty-five children successfully used one or other of these; but fifty children obtained the correct answer by one of the fourteen methods, which had been developed without being taught.

Somewhat similar results are quoted by McIntosh (1978), who describes a variety of (presumably) self-devised methods, which depend on conceptual understanding and are reasonably efficient, at least for the individual concerned. A particularly large range of different procedures is adopted for the subtraction

$$431 - 145$$

An example is:

'Took 1 from 431. Took 45 away from 430 made 385. Then added 1 to make 386. Finally I took 100 away, answer 286.'

The variety of thought and thought-processes is surprising and sometimes amusing, and contrasts strongly with the narrow path which the learning of standard techniques might be expected to provide.

An article by Dichmont (1977), also gives some illustrations of individually-developed operating procedures, in which numbers are split up, pushed around and combined with fluency and confidence.

92

A six-year-old boy was given the problem

$$27 + 5 =$$

He immediately replied '32', and when pressed for an explanation said:

'I take 3, put 3 onto the 7, ... which makes 10, that's 30, ... and another 2 left, makes 32.'

The most striking part of Dichmont's discussion is the element of strategy involved in this process: the effective solution of arithmetical problems involves a search for simplification rather than the immediate application of the technique considered relevant.

These results all suggest that the putting together of skill elements to produce an *ad hoc* method for the solution of a given problem is a more natural way of operating for children than the learning of a standard procedure. Combined with what we know of the difficulty of performing algorithms such as those for multiplication and division with speed and accuracy, we are forced to conclude that this is not a mode of functioning for which the human brain is well designed; in this respect it is unlike the calculator or the computer.

In general points made here are also exposed by Skemp (1976-1979) in his discussion of relational and instrumental understanding. He points out that it is possible to teach 'instrumental mathematics' (concentrating on skills) or 'relational mathematics' (developing conceptual understanding and structure). It is suggested that there are distinguishable groups of teachers following each of these practices and this raises two questions: 'Does this matter?' and 'Is one kind better than the other?'. In spite of the temptation to answer each of these with a simple 'Yes', Skemp goes on to suggest criteria for choosing, and advantages and disadvantages associated with each approach. Thus instrumental mathematics is said to be easier to learn but harder to remember. In instrumental mathematics the rewards may be more immediate, and the right answers appear quickly, but it is claimed that relational mathematics is more flexible and more adaptable to new tasks.

References for Chapter Five

ANDERSON, R.C., KULHAVY, R.W. and ANDRE, T. (1971): 'Feedback procedures in programmed instruction', *J Ed Psych*, 62, 148-56.

ANDERSON, R.C. KULHAVY, R.W. and ANDRE, T. (1972): 'Conditions under which feedback facilitates learning from programmed lessons', *J Ed Psych*, 63, 186-8.

ASHLOCK, R.B. and HERMAN, W.L. (1970): *Current Research in Elementary School Mathematics*. New York: Macmillan.

BROWNELL, W.A. and CHAZAL, C.B. (1935): 'The effects of premature drill in third-grade arithmetic', *J Ed Res*, 29, 17-28.

BROWNELL, W.A. and CARPER, D.V. (1943): *Learning the Multiplication Combinations*. Duke University Research Stud Ed (7), Durham, NC: Duke University Press. p.177.

BROWNELL, W.A. and MOSER, H.E. (1949): *Meaningful versus Mechanical Learning: a Study in Grade II Subtraction*. Duke University Research Stud Ed (8), Durham, NC: Duke University Press. p.207.

COSGROVE, G.E. (1957): The effect on sixth-grade pupils' skill in compound subtraction when they experience a new procedure for performing this skill. Doctor's dissertation, Boston University School of Education.

DICHMONT, J. (1977): 'Solving our own problems', *Times Ed Supp.*, 18th March, p.41.

GINSBURG, H. (1977): *Children's Arithmetic*. New York: Van Nostrand.

GRUNDIN, H.U. (1969): 'Response mode and information about correct answers in programmed instruction', In: MANN, A.P. and BRUNSTROM, C.K. (Eds): *Aspects of Educational Technology*, Vol. III. Pitman.

HART, K. (1981): *Children's Understanding of Mathematics (11-16)*. London: Murray.

HARTLEY, J.R. (1980): Using the computer to study and assist the learning of mathematics. Paper presented to the BSPLM, University of Nottingham, January.

JONES, D.A. (1975): 'Don't just mark the answer - have a look at the method', *Maths in School*, May, 29-31.

McINTOSH, A. (1978): 'Some subtractions: what do you think you are doing?', *Maths Teaching*.

NITSCH, K.E. (1977): Structuring decontextualized forms of knowledge. Ph.D. dissertation, Vanderbilt University, Nashville, Tennessee, USA.

PLUNKETT, S. (1979): 'Decomposition and all that rot', *Maths in School,* May, 2-5.

RESNICK, L.B. (1980): 'The role of invention in the development of mathematical competence'. In: KLUWE, R. and SPADA, H. (Eds): *Developmental Models of Thinking.* Academic Press.

SKEMP, R.R. (1976): 'Relational understanding and instrumental understanding', *Maths Teaching,* December, 20-6.

SKEMP, R.R. (1979): *Intelligence, Learning and Action.* load. New York: Wiley.

STEFFE, L.P. and PARR, R.B. (1968). *The Development of the Concepts of Ratio and Fraction in the Fourth, Fifth and Sixth Years of the Elementary School.* University of Wisconsin, Madison.

TAIT, K., HARTLEY, J.R. and ANDERSON, R.C. (1973): 'Feedback procedures in Computer-Assisted Arithmetic Instruction', *B J Ed Psych,* 43, 161-71.

WILLIAMS, J.D. (1971): *Teaching Technique in Primary Mathematics.* NFER.

WOODS, P. and HARTLEY, J.R. (1971): 'Some learning models for arithmetic tasks and their use in computer-based learning', *B J Ed Psych,* 41(1), 35-48.

Chapter Six
Children's Understanding of Concepts
in Specific Topic Areas

1. INTRODUCTION

This chapter concerns research on the understanding of
particular topics of the mathematics curriculum. Much of
this research gained its inspiration from Piaget's early
studies of child's acquisition of number, space and geometry.
The main aim of these in Piaget's own work was to establish
general stages of development, and this influence can be
seen in much of the later work. However, the point of greatest interest for the teacher of mathematics is in what is
revealed about the children's own thinking in relation to
these important areas of mathematical knowledge. The
Piagetian method of asking a child probing questions about
a carefully chosen situation has proved very powerful in
revealing aspects of the child's thinking about which the
teacher needs to know if he is to help him successfully to
improve his mathematical understanding. There is now a

substantial body of such knowledge and its relevance for
the mathematics teacher is abundantly clear.

Although the detailed results themselves are of the
greatest interest, it is also relevant to consider what general conclusions may be reached by reflection on this material. One might first conclude that we have here a great
many difficulties which teachers would not have suspected,
for example the loss in facility of some 20 per cent in the
move from vertical to horizontal presentation of addition
and subtraction items, and similarly in the change from
$84 \div 4$ to $84/4$. Then the fact that, while nearly 70 per cent
of English 11-year-olds can answer correctly how many gallons
of petrol will be needed by a lorry to do 108 miles at 9
miles per gallon, only 38 per cent can answer the question
which says that if there are 25 cars noted to every 3 buses
in a traffic count, and 12 buses pass in one hour, how many
cars can be expected to pass in an hour, may cause a little
surprise, too. Or to take a third example, $e + f = 8$,
$e + f + g = $? was answered correctly by only 41 per cent of
third and fourth year secondary school children. Such
results can be found on almost every page of this chapter
and they serve to remind us that we have so far only begun
to scratch the surface of understanding just what it is that
makes mathematical tasks difficult for children. Some
principles have indeed been teased out from the mass of data,
and it is insofar as this knowledge becomes further increased
and further disseminated amongst teachers that we may hope
for a real and steady increase in the effectiveness of
mathematics teaching.

Another feature which may strike the reader is the
relatively small increases from year to year which appear
when general tests of mathematical understanding (such as
those developed by CSMS) are set to large groups of pupils.
It is clear that a fairly large amount of mathematical
teaching and practice leads to a fairly small increase in
the understanding of fundamental ideas. How far do we teach
with the awareness of this fact, that is, that most of the
detail of what we are teaching is going to be lost and only
the general ideas retained? Could teaching be more effective
if we diagnosed in more detail the particular difficulties
which individual pupils needed to overcome, rather than
offering teaching at large which may or may not suit the
needs of individuals? Or is a solution to be sought in the
direction of offering, as teaching material, situations from
which different pupils are able to take the different things
which they need at the time? Or should we look more carefully at the levels of mastery which we require pupils to
reach before passing to a new topic? However these questions
may be answered, and they all need further study, the
material in this chapter should enable teachers more easily
to diagnose the difficulties their pupils have in many areas
of the curriculum.

The investigations described in this chapter are of
several types. At the most basic level, they attempt to
specify the relative difficulty of certain tasks or questions;
this may involve quoting a facility - the percentage of res-
pondents correctly able to complete the task. But beyond
this, the levels of difficulty may be seen to represent a
pattern of development; children proceed from being able to
perform only the easiest tasks to being able to handle an
increasingly wide range of questions. There are also teaching studies which seek to compare treatments, using perhaps
different embodiments or different orderings of the conceptual structure, or different styles of teaching. Where
these have results which relate to the learning of a particular topic, such results will be included here.

The structure of this chapter follows in general the
curriculum model adopted for the APU surveys.

2. NUMBER

We may consider this field of learning as beginning with the
notions of sorting and comparing and ordering, and the
beginnings of the acquisition of number in the early years
of schooling. There follows an extensive body of knowledge
concerning the meaning of the numeration system, and the
four operations of addition, subtraction, multiplication and
division, and the many relationships involved in understanding and applying these operations. The knowledge of
particular number facts, and the development of skills of
computation, represent another major part of this field.
Fractions, percentages, ratio and proportion form another
major area; so do directed numbers. Alongside the understanding of the nature of these numbers and their internal
relationships goes the ability to recognize them in appropriate situations, which are usually met in the curriculum
in the form of verbal problems. In the APU classification
these are described as applications. We shall attempt to
indicate what research exists, as well as what gaps exist in
our knowledge in regard to the understanding of each of these
areas.

Early Concepts of Classification and Number

One of the chief contributions of research in this area, from
the point of view of the teacher, is the exposure of the
distinctions between, say, simple computational skills such as
knowing the simple addition facts, and an awareness of general
number properties such as the invariance of the number of
objects of a set under rearrangements; or between the ability
to recite the number names in the correct order, and the
awareness that when counting a set, the same final number is
reached irrespective of the order in which the objects are
taken. Much of this work has by now been absorbed into school
materials and teachers' guidelines such as, for example, the

ILEA Primary School Mathematics Guidelines and the associated
Checkpoints. The original source is Piaget's *The Child's
Conception of Number* (1952); an exposition written for
American student teachers of much of this work is Copeland:
How Children Learn Mathematics (1974). Perhaps the best
compromise which combines authority with accessibility is the
collection of lectures at the conference, *Piagetian Cognitive
Development Research* and *Mathematical Education* (edited
Rosskopf *et al.*, 1971). A more detailed discussion of the
acquisition of counting abilities is that by Gelman and
Gallistel (1978). Much of the research has focused on the
order of acquisition of different aspects of the conceptual
structure. One of the more recently exposed aspects of this
is that the key number awarenesses, such as the invariance
under rearrangement, are first appreciated for sets of quite
small numbers of objects, e.g., up to three, and only gradually extended to other small numbers, and finally to bigger
numbers outside the immediate range of concrete experience.
The development of number concepts through work with sets is
an established feature of current primary school practice;
the work is often developed into a more formal study of set
language and notation. Some examples of this kind of material
and the facilities achieved by 11-year-old children are
provided by the items from the APU Primary Survey given on
p. 100.

Items from sets and relations
Item cluster: sets
Mean sub-category score–51 per cent

Response analysis		Item facility		
Incorrect	5%	87%	N1	This circle marked A has five numbers in it and represents the set $A = (2, 8, 3, 7, 5)$
Omitted	8%			
a) Incorrect	15%	66%	N2	
Omitted	19%			
b) Incorrect	45%	22%		
Omitted	33%			
c) Incorrect	11%	66%		
Omitted	23%			
Incorrect	41%	53%	N3	
Omitted	6%			
Incorrect	44%	43%	N4	
Omitted	13%			

N1 A circle marked A has five numbers in it and represents the set $A = (2, 8, 3, 7, 5)$

A: 2, 3, 7, 8, 5

Two of the sets below are equal to set A. Tick them both.

B: 2, 3, 7, 8, 5
C: 2, 3, 4, 7, 8, 5
D: 1, 4, 6, 9, 0
E: 7, 2, 8, 3, 5

N2

E is the set of boys in our class.
F is the set of boys who play football.
C is the set of boys who play cricket.

a) Put B on the diagram to show Bill who plays both cricket and football.

b), c) Andrew is marked A on the diagram. Write down two things you can say about Andrew's sporting activities.

....

....

N3 A baker delivers bread to 90 customers.
60 customers take white bread and 40 take brown bread.
How many customers take both white and brown bread?

N4

R is the set of red shapes.
L is the set of large shapes.
T is the set of triangles.

Put a △ in the space where a large, red triangle would be.

We now return to the discussion of recent research concerning young children's understanding of number.

A logical distinction can be made between ordinal and cardinal properties of number: in the first category, we are concerned with the order of the numbers in a sequence, and the properties of the order relationship; and in its cardinal form, we recognize number as the property which two sets share because they can be put into one-one correspondence, regardless of the nature or position of their elements. Both these aspects are also important psychologically (see Lovell, 1978), and experiences can be provided in which the attention of young children is drawn to either kind of property, or to both. Piaget and Szeminska (1952) have shown that pre-school children have difficulty in constructing ordered sequences (for which it is necessary to recognize that a given number is simultaneously less than one immediate neighbour and greater than the other); also, they do not always recognize that the numerical equivalence of two sets is independent of the kind of elements and the way they are laid out. Piaget and Szeminska maintain that the ordinal and cardinal aspects of number develop concurrently, and that when these become systematic or 'operational' they become integrated into an over-all number-concept; Brainerd, on the other hand (1979) maintains that the concept of 'ordinality' is acquired earlier than the concept of 'cardinality'. However, his tests are so disparate that it is not clear why the abilities required could in any way be thought of as equivalent: the tests of ordinality in particular are in the contexts of length and weight. Thus, children were asked to decide which of two sticks (glued to a board to prevent direct comparison) was larger, by comparing each in turn with another stick (which turned out to be larger than one and smaller than the other). This task requires use of the transitive property of the order relation; and Brainerd's results suggest that this property is recognized by children before that of 'cardinality' in the sense of numerical equivalence of two sets. A replication of this finding is provided by Gonchar and Hooper (1975); but it is not clear whether the concept being tested by Brainerd or Gonchar and Hooper is equivalent to 'ordinal number' - and it is certainly context-related.

The development of cardinal concepts seems to have been studied more thoroughly than ordinal concepts, perhaps because the latter have usually been investigated in the context of length rather than number in an attempt to control the cardinal dimension. The best known studies of cardinality are Piaget's number-conservation tasks, in which children are asked to compare two arrays made up of the same number of elements, after one of these arrays has been transformed. Because these tasks have become so familiar it is easy to forget their rationale, which is that children do not simply 'have' or 'not have' a concept such as 'number' but rather that a concept can be understood to different degrees, and that the

nature of this understanding, and its stability, can best be
revealed by devising situations where this understanding
breaks down. However, it sometimes appears that Piaget has
forgotten this principle himself, as he gives the impression
that children only have a 'proper understanding' when they can
succeed on his particular tasks.

According to Piaget, this success occurs when the child's
understanding has become fully 'operational', by which he
means that the child is no longer distracted by isolated
perceptual cues (e.g., length or density), but instead
recognizes that the array still contains the same elements
(identity) and that the transformation can be reversed
(reversibility); moreover, the child is now able to
co-ordinate the perceptual cues, in that, for example, an
apparent increase in number due to an increase in length is
compensated by a decrease in density. This compensation
argument is of particular importance, since a child who is
unable to reconcile the apparent unreliability of the
perceptual cues would seem to have only a limited understanding of conservation. However, this does not mean that
it is only through learning to recognize this compensation
that conservation develops. It is possible for children to
learn to ignore some perceptual cues, giving accurate
responses, without necessarily understanding why a certain cue
is appropriate in a particular situation. For example,
Bryant (1974) has shown that children can be trained to give
conservation responses well below the age at which Piaget
says number-conservation becomes operational (about seven
years), by presenting them with numerous instances in which
cues such as length give conflicting judgments. Gelman
(1969) has produced similar results, and she has also
(Gelman, 1972) shown that even without training, some children
as young as three or four years can ignore misleading cues if
the numbers are sufficiently small (two or three elements, as
opposed to seven, say, in the original Piagetian task.) This
result fits with an interesting study by Young and McPherson
(1976), who investigated the ability to compare (unequal)
numerical quantities in the face of misleading perceptual
cues, for numbers which children could subitise (recognize
'instantly'), or count, or which were too large to count.
It can be seen from their results, below, that children are
most likely to ignore misleading cues if the numbers are
familiar or easily counted, and that as they get older they
can extend this facility to numbers whose absolute value can
not be assessed immediately. This improvement may in part be
due to an accumulation of experience, but along with this
some qualitative difference in thinking develops, which
enables children to co-ordinate different cues and to make
sense of the apparent conflict.

	Average subitizing level	Percentage of trials on which invariance shown		
		Numbers within subitizing range	Numbers outside subitizing range but within counting range	Large numbers and counting prevented
Age 5 (N=20)	4.35	48	32	27
Age 6 (N=20)	4.60	94	83	76

Conservation of number, particularly in the sense of appreciating that certain transformations leave the number of elements in an array unchanged, is clearly an important aspect of number, and one which may not be fully grasped, even by children who can count the elements concerned (and perhaps even know some simple number facts such as 6 + 2 = 8). This means that we need to identify separately certain aspects of number competence, particularly

    (a)   the appreciation of the order properties,
    (b)   invariance of number under some transformations, and
    (c)   the ability to count.

It does not, however, mean that counting and other number skills should not be practised until the child is deemed to be 'operational' on 'conceptual' tasks such as those of Piaget. Rather, it seems likely that traditional skills like counting, and reading and writing symbols, can enhance the understanding of order and conservation, provided children can be encouraged to recognize and reflect upon these aspects of number. Conversely, such understanding supports and reinforces the skills.

A considerable amount of work is currently in progress in the United States on counting and early number. Fuson and Mierkiewicz (1980) have made a detailed analysis of the act of counting, and follow-up work is in progress to study how this is extended to the counting of larger numbers, e.g., up to 100. Carpenter and Moser have made a longitudinal study of development of children's ability to solve verbally-expressed addition and subtraction problems, both before and after instruction in addition (Carpenter and Moser 1979, Moser 1979, 1980, Carpenter 1980). Resnick and others have shown by the study of reaction times that from quite a young age, children, when asked to add two small numbers, say 3 and 5, begin from the larger and then count on the required number of additional steps. The time taken to respond is linearly related to the size of the smaller number to be added. This happens even if the commutativity of addition has not been pointed out. Similar studies of reaction times for questions requiring the products of numbers also show linear relationships, which suggest that the main links between these products in the memory may lie along the lines of a table of multiples (Resnick and Groen 1975).

Extensive studies have also been made of the frequency of
different kinds of error in subtraction computations. Of
these the commonest is the subtraction of the smaller number
from the larger in a given column without regard to which
number is in the top row. Next in frequency come errors
concerning zeros, then errors concerning borrowing with zeros
(Brown and Burton 1978, Young and O'Shea 1981). Current work
by Resnick is directed at the teaching of the correspondence
between subtraction using Dienes blocks and the written
algorithm.

*The numeration system - place value (whole numbers and
decimals)*

The APU surveys provide two clusters of items which show
performance of English pupils at age 11 in this conceptual
area, given on pp.105-6.

Items from concepts (decimals and fractions)
Item cluster: decimals
Mean sub-category score—48 per cent.

Response analysis		Item facility	
Incorrect	19%	79%	F1 Put these decimals in order of size, smallest first. 0.3 0.1 0.7 0.6 . . . .
Omitted	2%		
0.56 is:		54%	F2 Tick the line that is correct. 0.56 is less than 1.3 0.56 is greater than 1.3 0.56 is equal to 1.3
greater than 1.3	29%		
equal to 1.3	11%		
Other incorrect	1%		
Omitted	5%		
Incorrect	41%	47%	F3 How many times is 0.1 greater than 0.01? . . . .
Omitted	12%		
Incorrect	54%	34%	F4 What number is 10 times 0.5? . . . .
Omitted	12%		
Incorrect	76%	21%	F5 Put these decimals in order of size, *smallest* first. 0.07 0.23 0.1 . . . .
Omitted	3%		

Items from concepts (whole numbers)
Item cluster: numeration
Mean sub-category score–59 per cent.

Response analysis		Item facility	
Incorrect 107	1%	86%	E1 Put a ring around the number which is the same as 7 tens. 107 70 7 710
Incorrect 7	2%		
Incorrect 710	6%		
Other incorrect	3%		
Omitted	2%		
Incorrect	24%	71%	E2 Which number is ten times 100? . . . .
Omitted	5%		
Incorrect	27%	64%	E3 The number which is one less than 2010 is. . . . . .
Omitted	9%		
275 752 725 572	4%	54%	E4 Put a ring around the line which is in order of size. 275 752 725 572 257 527 275 752 572 752 527 275 257 275 725 752 572 527 257 275
257 527 275 752	4%		
572 752 527 275	5%		
572 527 257 275	8%		
Other incorrect	11%		
Omitted	14%		

The report also states that only 21 per cent of pupils could apply the general concept of place value in the new context of an unconventional number system. More detailed information about the understanding of concepts of decimals and fractions is contained in the report of the CSMS research (Hart, 1981).

Children's grasp of the working of the place value
system is to some extent tested by their ability to add or
subtract one from a number at the critical stage when a zero
appears in the units column. The following item is from the
APU Primary Survey:

'The number which is one less than 2010 is ....?'

The quoted facility (at age 11) is 64 per cent. Following on
from this, the CSMS test includes this question:

'This meter counts the people going into a football stand.

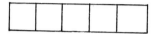

After <u>one</u> more person has gone in, it will read:

The facilities are:

Age: 12   13   14   15
                  68% 77% 86% 88%

Placing whole numbers in order of size has a broadly similar
level of difficulty: an item requiring this was answered
correctly by 55 per cent of 11-year-olds and 80 per cent of
15-year-olds in the survey.

These figures give some clue about the development of
children's understanding of place value, suggesting that this
understanding is achieved in general terms by a slight
majority of 11-year-olds, and identifying the size of the
minority who still do not understand the notation at age 15.
The grasp of place value is, however, often very tenuous.
Questions containing numbers over a thousand, especially if
these involve a number of zero place-holders, are more
difficult and expose children's uncertainties. These are from
CSMS:

'Write in *figures:*   Age   12   13   14   15
    Four hundred thousand   Facility 42   51   57   57%
    and seventy three.'

'5214   The 2 stands
  ↑   for 2 hundreds

'521400   The 2 stands   Age   12   13   14   15
  ↑   for 2 ....'   Facility 22   32   31   43%

We might compare this last item with the facility of 70 per
cent with which 15-year-olds are able to correctly specify the
value of the digit 7 in 1728, in the APU Secondary Survey.

In a similar way, the facility is reduced when digits in decimal places are to be evaluated. The Secondary Survey quotes a facility of 40 per cent for the item:

'What is the value of the figure 1 in the decimal 2.31?'

In questions of this nature, handling only one place of decimals appears to be relatively easy: two or more places produce a sharp increase in difficulty. This is reflected by the allocation to levels in the CSMS work, where one level is named 'decimal-tenths', and 'hundredths, thousandths, etc.' are assigned to a higher level. The distinction also appears clearly in two items in the APU Primary Survey. 80 per cent of 11-year-old pupils were able to put decimals written to one place in order of size, but the introduction of a second place reduced the facility to 21 per cent. The main difficulty is to recognize, for example, that 0.4 is *greater* than 0.26, though 4 is less than 26.

In summarizing this section, two features are perhaps worth comment. One is the considerable difference in the age at which understanding is acquired. Many 11-year-old children show a better grasp of the notation than a sizeable minority of 15-year-olds. The other remarkable feature is the tenuous nature of this understanding, and its instability in more complicated situations: the area which is perhaps particularly weaker than might be expected is in whole numbers over a thousand.

Numerical operations and their relationships

Under this heading the APU Primary Survey tested the understanding of prime numbers, odd and even, factors and multiples, the understanding of commutative, associative and distributive laws, and so on. Their summary report says that the majority of pupils were successful with items involving place value and the general laws, but only a smaller proportion could deal with more formal ideas such as those of factor, prime number and negative number. Little further detail is given in the report of the first survey, but it may be that more detailed information is published in subsequent years. An item in the Secondary Survey tested 15-year-old pupils' awareness that addition and multiplication are associative whereas division is not, by asking whether a number of statements such as $(32 \div 6) \div 15 = 32 \div (6 \div 15)$ were true. About 40 per cent appear to appreciate this point; it should be noted that formal knowledge of the terms designating the properties was not necessary to answer these items correctly.

It is to be hoped that the APU will in fact release a wider range of items in this area, since there seems to be a lack of research here. Studies which will be described below identify levels of difficulty in the understanding of the

108

general algebraic laws (commutativity, and, so on), between different kinds of subtraction (e.g. pure subtraction, complementary addition, comparison), and between subtractions as applied to discrete objects overlapping distances, or displacements on a number line. It would be helpful to have more detailed knowledge of pupils' understanding of, for example, the relation between multiplication and division, the various relationships occuring within the multiplication table, and the whole mass of relationships underlying computational procedures, both standard and informal.

Lovell (1972) has reported experiments showing that many six to seven-year-olds are able to memorize number facts employing symbols, e.g., 6 + 2 = 8, but are unable to recognize the concept of addition in a physical situation: this is another aspect of the separateness of different types of mathematical knowledge. Other work (Brown, 1969), using concrete materials and small numbers with pupils of average ability, showed that the closure of addition and multiplication were understood around the age of seven, the properties of nought and one by eight, commutativity and associativity by nine, and distributivity by eleven years of age; but pupils' performance could be advanced or retarded considerably compared with this norm (see Lovell, 1978). Studies by Lunzer, Bell and Shiu (1976), focusing on the understanding of subtraction and the development of the use of negative numbers, showed considerable differences in the level of difficulty of situations which demanded subtraction. Those which could be described as pure subtraction (taking away) were the easiest, and complementary addition situations were the hardest, with complementary subtraction situations intermediate in difficulty. The differences were fairly marked with seven-year-old children but less so with nine-year-olds. Through the age range 7-13 there were marked differences in difficulty according to the kind of structure in which the subtraction had to be done. Where the numbers referred to discrete objects the youngest children could perform satisfactorily; next in difficulty were situations where the numbers referred to overlapping distances or displacements on a number line. The most difficult situations were those where two displacements had to be combined to give a third, without there being a designated starting-point or a numbering. (This situation was realized by marking the different points on the line with coloured stickers.) Both a straight line and a six-point clock face were used in this final case, but the cyclic aspects of this second situation appeared to present relatively little difficulty compared with the other aspects mentioned. These more difficult tasks were satisfactorily performed by 13-year-old children. Another experiment by the same team, using 3-term expressions of the form 8 - 13 + 26, showed a strong tendency in such items to commute the subtraction (taking 8 from 13), right up to the fourth form of the secondary school; this suggested that, though directed numbers had been taught, they were not understood sufficiently deeply to be used in this situation.

Recognition of numerical operations

This is becoming increasingly recognized as an important, and by no means easy, aspect of the understanding required to use numbers in real situations. The APU Primary Survey provides a cluster of items concerning applications of division:

Items from applications of number

Item cluster: division
Mean sub-category score–54 per cent.

Response analysis		Item facility	
Incorrect	11%	83%	J1 A bar of chocolate can be broken into 18 squares. There are 6 squares in each row. How many rows are there? . . . . rows
Omitted	6%		
Incorrect	17%	76%	J2 Spoons are sold in boxes containing half a dozen. If I want 30 spoons, how many boxes shall I buy? . . . .
Omitted	7%		
Incorrect	25%	66%	J3 150 people are coming to see the school play. The chairs are arranged in rows of 15. How many rows of chairs will be needed? . . . .
Omitted	9%		
Incorrect	34%	53%	J4 256 children are going to have tea at the Christmas party. 8 children can sit at a table. How many tables will be needed? . . . .
Omitted	13%		
Incorrect	37%	25%	J5 A batting average in cricket is found by dividing the number of runs scored by number of times out. Fill in the following table. Name No. times out No. runs scored Average Boycott 5 500 . . . . . .
Omitted	38%		

This invites speculation regarding whether the increasing difficulty is a result of the differences in context or simply in the differences in size of the numbers involved (apart from the last item). We know in fact from other research that both factors are important (e.g. Hart, 1981). Studies by Rees at Brunel University, based on mathematical papers set to craft students in evening courses, have identified a number of items of high difficulty where again the recognition of the appropriate calculation appears to be the predominating factor, since corresponding numerical items in the same test are considerably easier (Rees, 1973).

A lot of literature exists concerned with the factors which make it more or less difficult to identify the correct numerical operation in a word problem. Most studies identify some variable which contributes to this: some have examined a number of variables in attempting to predict relative difficulty by multiple-regression analysis. One study by Jerman and Mirman (1974) considers 73 linguistic variables and a number of computational variables in this way.

There are two recent articles by Nesher (1976) and Caldwell and Goldin (1979) which perhaps deserve more detailed attention in that they both attempt to summarize the conclusions of earlier work as well as investigate these by further experiment. Caldwell and Goldin identify a number of variables which significantly affect word-problem difficulty (see list of references in 1979 paper).

The study by Nesher (1976) investigates the effect of four such variables: context, superfluous information, number of steps, verbal cues. Thus, she tested children on a set of questions in each of four contexts; and in each of these contexts the questions were designed to include one or two steps, superfluous information, verbal cues, and combinations of these. The set of questions for one of the contexts is shown in the table below:

Question 2. A division of Cloth for Equal Suits.

2   A tailor sewed out of a 56 m long piece of material 7 identical suits with a modern and handsome cut. Find out how long was the material required for each suit.

2A A tailor divided up a 56 m long piece of material for 7 identical suits with a modern and handsome cut. Find out how long was the material required for each suit.

2B A tailor sewed out of a 56 m long and 80 cm wide piece of material 7 identical suits with a modern and handsome cut. How long was the material required for each suit?

111

2C A tailor divided up a 56 m long and 80 cm wide piece
    of material for 7 identical suits with a modern and
    handsome cut. How long was the material required for
    each suit?

2D A tailor sewed out of a 60 m long piece of material 7
    identical suits and had a 4 m long piece of material
    left. Find out how long was the material required
    for each suit.

2E A tailor divided up a 60 m long piece of material for
    7 identical suits and had a 4 m long piece of material
    left. Find out how long was the material required for
    each suit.

2F A tailor sewed out of a 60 m long and 80 cm wide piece
    of material 7 identical suits and had a 4 m long piece
    of material left. How long was the material required
    for each suit?

2G A tailor divided up a 60 m long and 80 cm wide piece
    of material for 7 identical suits and had a 4 m long
    piece of material left. How long was the material
    required for each suit?

This context turned out to be the most difficult of four that
Nesher used, with 'A division of wooden boards into equal
shelves' the easiest. The average difference in facility is
quite large, compared to the facilities themselves. The
results quoted show that the presence of an extra step adds
to the difficulty (e.g., item 2D compared to item 2), as does
the presence of superfluous information (e.g., item 2B
compared to item 2). Nesher also found a significant interaction effect of context and superfluous information: the
presence of superfluous information matters much more in less
familiar contexts.

In this study, the presence of verbal cues does not have
a significant effect, to some extent conflicting with the work
of Jerman and Rees (1972) and an earlier paper by Nesher and
Teubal (1975) which emphasise the impact of verbal cues in
problem solving. However, Nesher's (1976) study compares
verbal cues with neutral words, whereas the Nesher and
Teubal investigations match reliable verbal cues against
'distractors', so that it may well be the distractors which
require special attention. A more detailed look at the
Jerman and Rees paper appears to confirm that it is the
distractors which produce the significant (adverse) effect
rather than the verbal cues.

A rather different approach is adopted by Caldwell and
Golding (1979), who set out to compare the relative difficulty
of four types of word problem: abstract factual (AF), abstract
hypothetical (AH), concrete factual (CF) and concrete

112

hypothetical (CH). Children were tested on 5 versions of each of the problem-types shown below, and it can be seen from the percentages in the accompanying table that the concrete problems are markedly easier than the abstract ones. This is still true when attention is restricted to the children who are definitely capable of the appropriate computations. It is of course probably true that the (primary school age) children tested had had less experience of abstract problems than concrete ones, and a substantial number of abstract problems were correctly solved by some young children.

		Proportion of item-types solved correctly (144 11-year-olds)
1. There is a certain given number. Three more than twice this given number is equal to 15. What is the value of the given number? (Abstract factual. No change is described.)	AF	39%
2. There is a certain number. If this number were 4 more than twice as large, it would be equal to 18. What is the number? (Abstract hypothetical. The number is not really 4 more than twice as large.)	AH	48%
3. Susan has some dolls. Jane has 5 more than twice as many, so she has 17 dolls. How many dolls does Susan have? (Concrete factual. No change is described.)	CF	58%
4. Susan has some dolls. If she had 4 more than twice as many, she would have 14 dolls. How many does Susan really have? (Concrete hypothetical. Susan does not really have 4 more than twice as many dolls.)	CH	57%

For concrete problems there is very little difference in difficulty between factual and hypothetical versions; but for abstract problems the hypothetical versions show up as consistently less difficult than the factual ones. It may be that the hypothetical phrasing of the abstract problem suggests the appropriate mathematical operation more clearly than the factual phrasing.

From all these studies, it is possible to identify some five categories of variable which affect the level of difficulty of a problem as follows:

1. *Context*. Brownell and Stretch's early work (1931) on the effect of familiar and unfamiliar contexts is confirmed by later investigations. Some indication of the amount of variation involved is provided in the APU Primary Survey (1980), where the facility of questions involving simple division of whole numbers varies from 83 per cent to 25 per cent. (See page 110).

2. *Readability*. Used in its widest sense, the term 'readability' covers a whole range of linguistic factors (actual time needed to read problem, length of sentences, complexity of sentence structure) which affect the difficulty of questions. It is not particularly startling, however, to discover that if a problem is more difficult to read then it is more difficult. Much of the work in this area, particularly that by Jerman and others (1972, 1974) is concerned with identifying more precisely the features which make problems more difficult in this way: the 73 linguistic factors of Jerman and Mirman (1974) take this to extremes. It is perhaps inappropriate to provide a detailed analysis of this work here, except to comment that some linguistic forms are obviously simpler than others. This aspect is considered again in Chapter Eleven.

3. *Size and Complexity of Numbers*. The striking feature of this particular effect is that complicated numbers make it more difficult to *recognize* the operation or relationship involved in a word problem: it is not just a question of the computational element being harder. There is quite a lot of evidence of this in the CSMS tests, where increasing the size of the (whole numbers) in a problem which remains parallel in other respects reduces the facility by anything up to 36 per cent with a mean reduction of about 14 per cent. Presenting a problem of the same type with decimals makes it considerably more difficult, particularly in some cases seen as awkward (e.g., multiplying by a number less than one, dividing a smaller by a larger number) where correct response rates dropped to under 20 per cent.

4. *Number and Type of Operations and Stages*. There is a lot of work in this field, some of which was quoted earlier in this section. In the CSMS report on the Number Operations test, the order of increasing difficulty of recognition is given (in general terms) as +, -, ÷, x. This, however, needs some qualification because of the particular forms of the operations quoted.

Thus, for multiplication especially, the 'cartesian product' form is much harder (facility about 27 per cent lower) than a 'repeated addition' type of multiplication: examples quoted are:

A shop makes sandwiches. You can choose from 3 sorts of bread and 6 sorts of filling.

How do you work out how many different sandwiches you could choose?

A bucket holds 8 litres of water. 4 buckets of water are emptied into a bath.

How do you work out how many litres of water are in the bath?

Nevertheless, when children were free to choose the form of the operations, by being asked to 'write stories' for expressions like 9 x 3, the order +, -, ÷, x was confirmed.

Nesher's (1976) study provides confirmation that the inclusion of more than one operation and more than one step increases the level of difficulty.

5. *Distractors*. Many forms of distractor can be introduced (unwittingly or otherwise) into word problems. The presence of superfluous information has been found to affect the level of difficulty, though the effect is somewhat dependent on the familiarity of the context. A search for verbal cues, which has been offered as a facilitating procedure by some, is frequently not appropriate. Reliable cues sometimes exist, but words out of context can just as easily act as distractors.

It is not always obvious which variable is creating the difficulty. The following example is provided by Rommetreit (1978). With an array like this some children who failed to

correctly identify

'the one of the white circles that is second largest',

succeeded when asked to identify

'the one of the snowballs that is second largest'.

It might appear that this is a difference in context: it is reminiscent of Caldwell and Goldin's distinction between abstract and concrete situations. However, the first question is linguistically more complex than the second: the child has to decide whether the 'second largest' is to refer

to the whole phrase 'white circles' or just to 'circles'.
Perhaps it is this which determines most of the difficulty.
Some kind of reformulation of the questioning is needed to
discover just what is happening.

Some recent (unpublished) work of a preliminary nature
by Swan (1980) underlines the difficulties by pointing out
the errors made in choosing the correct operation. An
unwillingness to divide smaller numbers by larger is related
to an apparent unawareness of the non-commutative nature of
division:

> The fuel tank of a Mini holds 5.5 gallons of petrol.
> How many litres does the tank hold?
>
> (Note:- 1 litre = 0.22 gallons)

A pair of 5th year pupils are discussing the item

Boy: So, in this it's already told you what one gallon is, so .....

Girl: 0.22 gallons *into* 5.5 gallons.

Boy: That will give you the number of litres.

Int: (points to the choice of answers beneath the problem)

Which of these answers is doing that?

Girl: (g). (0.22 ÷ 5.5)

The misconception that 'multiplication always makes
things bigger' is also found to be prevalent:

> In a supermarket, an old lady notices that a packet of
> pork chops weighing 1.07 lbs is priced at exactly £1.
> If she has only £0.86 in her purse, what is the maximum
> weight of chops that she can buy?

In the following extract, two 5th year girls are
considering the 'supermarket' item. Initially, they
were unable to choose an operation to perform, so the
Interviewer suggested replacing 1.07 by 4 lbs and
replacing £0.86 by £3.

G1, G2: She can buy 12 lb. Multiply 3 by 4. (an immediate, simultaneous answer)

Int: Why is that so obvious?

G1: Because it's 4 lb to the £1.

G2: So £1 is 4 lb in weight, so you multiply the 3 by the 4 to get 12.

Int: If I put the original numbers back in, will it still be multiply?

G1:   No, because you've got a less figure in your
        purse, so obviously you've got to *take away*
        pounds in weight because it's going to be a
        much lower answer in pounds in weight ... you've
        got less money in your purse, so you can't buy
        much ... obviously!

Understanding the nature of numerical operations in such
a way that these can be recognized in a variety of situations
is at least as fundamental as the ability to perform the
corresponding calculations. In practical terms, it is
arguably more important, and the increased availability of
calculators, for example, is of little value without this kind
of recognition. There are clearly identifiable difficulties
and some prevalent misconceptions in this area; and explicit
teaching appears to be needed if pupils are to acquire the
necessary competence.

Computation with whole numbers and decimals

The APU Primary Survey shows the displayed cluster but is
otherwise uninformative, except to say that the majority of
pupils were successful with addition and subtraction of whole
numbers but fewer pupils obtained correct answers for multiplication and division, whereas with decimals, addition and
multiplication were the easiest and division the hardest. A
study by Hitch (1978) therefore provides the most accessible
detailed information about computational performance, though
his work was done mainly with evening class students and not
with a representative population sample. More recent work
has attempted to identify the particular errors which
commonly appear. Subtraction has been examined this way in
intensive studies by Brown and Burton (1978) and by Young and
O'Shea (1981). (See Figure on page 118).

Items from computation (whole numbers and decimals)
Item cluster* naturals, multiplication
Mean sub-category score—57 per cent.

Response analysis		Item facility	
Incorrect	10%	84%	**G1** $50 \times 2 = \ldots\ldots\ldots$
Omitted	6%		
Incorrect	34%	57%	**G2** $76 \times 7 = \ldots\ldots\ldots$
Omitted	9%		
Incorrect	36%	49%	**G3** $381 \times 11 = \ldots\ldots\ldots$
Omitted	15%		
Incorrect	47%	38%	**G4** $124 \times 25 = \ldots\ldots\ldots$
Omitted	15%		
Incorrect	63%	29%	**G5** $\begin{array}{r} 314 \\ \times\,201 \\ \hline \end{array}$
Omitted	8%		

An attempt to specify stages in understanding fractions is described by Novillis (1976), who constructs a hierarchy of fraction concepts and investigates the extent to which some concepts depend upon others. In our Chapter Seven, Gagné's work on learning hierarchies will be studied in the context of developing teaching sequences: Novillis uses a similar procedure to try to identify conceptual dependencies. Novillis' work considers such ideas as associating a fraction with the area of a part of a shape (called a 'part-whole model'), with a sub-set of a larger set (a 'part-group model') or with a point on the number line: one of the stated conclusions is that the first two of these are both pre-requisite to the third.

Novillis describes the various models and subconcepts in some detail: some examples are shown below:

Level	Frame	Description
1	a.	Part-group, congruent parts—the student associates the fraction *a/b* with a set of *b* congruent objects, *a* of which are considered, or associates together two or more models representing this relationship.
		Example:
		3/4 of the objects are shaded.
	b.	Part-whole, congruent parts—the student associates the fraction *a/b* with a geometric region that has been separated into *b* congruent parts, *a* of which are considered, or associates together two or more models representing this relationship.
		Example:
		3/4 of the drawing is shaded.

Level	Frame	Description
2	a.	Part-group, noncongruent parts—The student associates the fraction *a/b* with a set of *b* noncongruent objects, *a* of which are considered, or associates together two or more models representing this relationship.

Example:

3/4 of the objects are shaded.

b. Part-group, comparison—the student associates the fraction *a/b* with the relative comparison of two sets A and B where $n(A) = a$ and $n(B) = b$ and all of the objects are congruent.

Example:

Set A   *Set B*

Set *A* is like 3/4 of Set *B*.

c. Number line—the student associates the fraction *a/b* with a point on the number line, where each unit segment has been separated into *b* equivalent line segments and the *a*th point to the right of 0 is considered, or associates together two or more number lines representing this relationship.

Example:

The point on the number line marked by *X* can be named 3/4.

d. Part-whole, comparison— the student associates the fraction *a/b* with the relative comparison of two geometric regions A and B, where the number of congruent parts in A is *a* and the number of congruent parts in B is *b* and the parts in Figures A and B are congruent.

Example:

Figure A   *Figure B*

Figure *A* is like 3/4 of Figure *b*.

e. Part-whole, noncongruent parts— the student associates the fraction *a/b* with a geometric region which has been separated into *b* parts that are equal in area but not congruent, *a* of which are considered, or associates together two or more models representing this relationship.

Example:

3/4 of the drawing is shaded.

The distinction between 'part' models and 'comparison' models may be worth noting. Steffe and Parr's (1968) study provides some comparison of the relative difficulty of part-whole (region) models and set-comparison (ratio) models. In some situations, one context is clearly easier than the other. Thus in the examples quoted earlier (in Chapter Five), corresponding to

$$\frac{4}{10} = \frac{\square}{5} \quad \text{and} \quad \frac{6}{15} = \frac{\square}{5},$$

the shaded-area model is easier. In 'missing denominator' examples, however, the ratio item is easier to operate (typically 60 per cent facility against 40 per cent for a sample of 11-year-olds). The underlying tasks are:

$$\frac{6}{18} = \frac{2}{\square} \quad \text{and} \quad \frac{6}{18} = \frac{3}{\square} :-$$

△ △ △
△ △ △   If there are six triangles   △ △
           for every fifteen circles,
○○○○○   for these two triangles there
           will be how many circles?
○○○○○
○○○○○

Six of sixteen equal parts are shaded in the top circle. In the bottom circle, into how many equal parts would you have to cut it so that if you shade three of these equal parts, the same amount will be shaded in both circles?

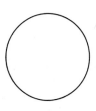

The result is interesting but should be interpreted cautiously. The first example is a set-comparison context, with elements shown distinct: in the second, the fraction is represented in a part-whole 'region' form. The comparison together with the distinctness of elements may make the identification of the pattern easier, but it is not known how this is interpreted. For example, it may still be the case that a part-whole region representation more readily conveys the fraction concept, and effectiveness may also depend on the way the region is shaded:

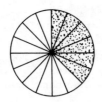

$\dfrac{6}{16}$

A number of studies of children's understanding of fractions have been undertaken in the United States, particularly at the University of Michigan, and a useful and substantial survey of the latter research is provided by Payne (1976). It is possible to see some identification of levels of understanding in this work, although much of it is concerned with the related task of finding effective teaching strategies.

Sometimes, this effectiveness is measured in terms of children's immediate subsequent performance on computational tasks. Carrying out calculations can enhance understanding, but an emphasis on the rules of manipulation without attention to the concepts involved is unlikely to lead to effective performance in the long term, and may also engender an unfortunate perception of mathematics as an arbitrary activity.

The idea of a fraction as an 'operator' is discussed by Hasemann (1980), who gave items from the CSMS Fractions test to German children who had been taught to regard fractions as operations. Thus 1/6 is considered as the operation of taking a sixth of something, whether a shape, a set or a number. Two of the items are shown below, and the order of difficulty is reversed for the German children:

(a) A part of the circle is dotted. Shade in 1/6 of the *dotted part*. What fraction of the *whole circle* have you shaded in?

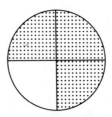

(b) Calculate: $\dfrac{1}{6} \cdot \dfrac{3}{4}$

		(a)	(b)
Percent correct	CSMS (14 years)	57	26
	German (12–15)	9 (30)	52

The items have in some sense a parallel structure, though a comparison is complicated by the fact that the second item can be solved using operators *or* by a rule; the first item requires *operators* but also *regions*, and though a reasonable proportion of Hasemann's sample could shade the correct area (30 per cent), it is noticeable that very few (9 per cent) could assign a correct fractional name to this region. This suggests that introducing fractions as regions provides a more meaningful notion of fractions, even if operators are more appropriate for some tasks.

There may be some temptation to move quickly to working with symbols alone in dealing with fractions, in that there are effective rules for solving problems expressed in this form. Without the conceptual framework, however, such skills are unlikely to be retained; in the CSMS Fractions test, about half the sample of 12-year-olds, but somewhat fewer older children could correctly solve $\frac{1}{3} + \frac{1}{4}$ and $\frac{1}{10} + \frac{2}{5}$ (as shown in the adjacent table)

	Percent Correct			
	12	13	14	15 year olds
$\frac{1}{3} + \frac{1}{4}$	54	38	35	45
$\frac{1}{10} + \frac{3}{5}$	55	38	49	45

Percentages are not conceptually different from fractions and ratios; it would be of interest to study the advantages of relating them more explicitly in teaching programmes. There seems to be little research in this area.

Ratio and proportion

The APU (Primary) cluster of released items is shown below (page 124).

The understanding of ratio and proportion is of course, related to fractions. In an interesting article, Van den Brink and Streefland (1979) described some situations in which young children (age 6-8) apparently know about ratio: they are all concerned with the relative sizes of the whole objects. For example, Coen (age seven)

'asks how big the propellor of a large ship is. His father tells him it would not fit into the woom. After a moment of silence he jumps to his feet saying: It is true. In my book on energy is a propellor like this (about 3 cm between his thumb and forefinger) with a little man like that (about 1 cm).'

A number of observations of this nature appear to show understanding of the concept of ratio in terms of physical size: children appear to have an operational understanding of ratio properties and similarity in their ability to interpret

123

Items from rate and ratio
Item cluster: proportionality
Mean sub-category score–39 per cent.

Response analysis		Item facility	
Incorrect	25%	69%	K1 A lorry uses 1 gallon of petrol every 9 miles. How many gallons would it use on a journey of 108 miles? . . . . gallons
Omitted	6%		
Incorrect	37%	57%	K2 In a game Jane got 3 points for every 8 points that Judy got, and Judy got 24 points. How many did Jane get? . . . .
Omitted	6%		
Incorrect	46%	51%	K3 A man can cycle a mile in 5 minutes and walk a mile in 20 minutes. How much time does he save when he cycles the 3 miles to work instead of walks? . . . . minutes
Omitted	3%		
Incorrect	50%	38%	K4 In a traffic count, there are on average 25 cars to every 3 buses. 12 buses go by in 1 hour, about how many cars would pass in one hour? . . . .
Omitted	12%		
Incorrect	60%	29%	K5 In the time that Zena takes to sharpen 2 pencils, Rachel can sharpen 3. When Rachel has sharpened 12 pencils, how many will Zena have done? . . . .
Omitted	11%		

124

what they see.

Freudenthal (1978) says:

'I go even as far as saying that congruence and similarity are built into the part of our central nervous system that processes our visual perception. The speed of identification of an object after the object itself or the observer has been rotated, or after its distance from the observer has been changed, presupposes, as it were, a computer program in the brain which eliminates this kind of transformation. While I do not understand at all what such a program looks like, its mere existence - which I do not doubt - is an enigma to me.'

Similarly, Van den Brink (1975, p.96) says ' ... it seems very likely that young children can take in and remember size ratios ... '.

Van den Brink and Streefland (1979) quote a particular instance in which this handling of ratio, associated as it is with the processing of perspective aspects of visual perception, leads to unwanted answers:

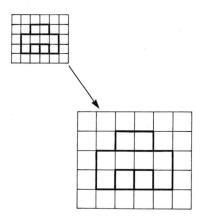

The lengths have been enlarged by a factor 2: what is the factor of enlargement of the area?

Many young pupils' answers suggest that they think the area has remained unchanged. 'The grid has grown', they reasoned, or 'we have come closer to the house'.

The early, visual appreciation of ratio properties is in strong contrast to the long-recognized difficulties which children in secondary school experience in handling *numerical* ratio problems. Piaget (1967) claims that a true understanding of ratio in such situations requires formal-level reasoning. Karplus (1974) has also identified a considerable lack of understanding of the ratio concept,

particularly in a task involving enlargement in the ratio
3:2. In work of this nature, a predominant error is the
adoption of an addition strategy in handling ratio problems:
children focus initially on the difference $a - b$ rather than
the proportion a/b.

Results from the CSMS Ratio test (Hart, 1978) suggest
that the most obvious, and probably most crucial, aspect of
ratio problems is the numbers that are involved. In a
representative sample of 14-year-olds (N = 767), nearly all
the children could cope with items requiring doubling or
halving (e.g., Item I, below) about half the children could
cope with items involving uneven multiples of a half (e.g.,
x 2½ or x 5/2, as in Item II, though children tended to solve
this item by building up to the answer, i.e., '25 is 10 plus
10 plus half of ten', which corresponds to 2 + 2 + 1,); very
few could cope when the multiplying factor involved fractions
other than multiples of a half (e.g., x 5/3 in Item III).

Item		Schematic Representation	Facility (14 year olds)

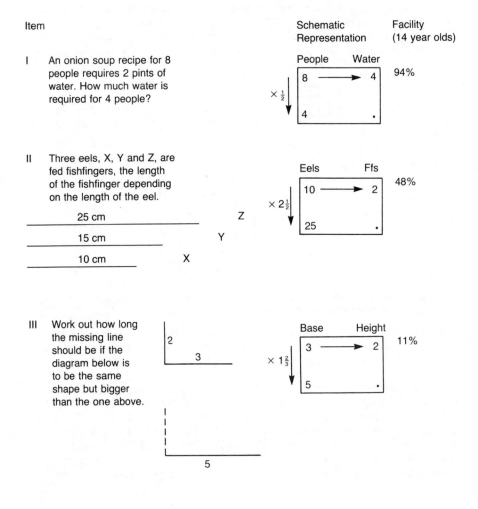

I   An onion soup recipe for 8 people requires 2 pints of water. How much water is required for 4 people?   94%

II   Three eels, X, Y and Z, are fed fishfingers, the length of the fishfinger depending on the length of the eel.   48%

25 cm Z

15 cm Y

10 cm X

III   Work out how long the missing line should be if the diagram below is to be the same shape but bigger than the one above.   11%

A closer examination of the data reveals that context is also an important determinant of item difficulty. Children seem to prefer to construct what Vergnaud (1979) calls 'scalar' rather than 'functional' relationships; e.g., in Item II children seem to focus on the relationship between the eels (10→25), rather than eels fish-fingers even though the latter is numerically simpler (10→2; 25 ?). This is supported by the fact that there is little difference in facility between Items II and II' (where the scalar relationships are of similar complexity and there is no simple functional relationship in II'.

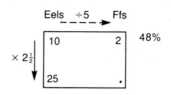

Item II

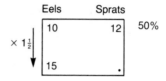

Item II

Another aspect of this dimension is the degree to which it is obvious from the context that the required relation involves multiplication (or repeated addition) rather than the addition of differences, say (the 'Addition Strategy', see below). For example, in the recipe question (Item I) it is fairly clear that the people will get *equal* shares of water, so that 4 people get half the water of 8 people (rather than, say, 4 pints less!). However, in a geometric context, like Item IV below, there is no such obvious correspondence - for example, one cannot add little K's together to make a larger K.

Item IV

The two letters are the same shape.
Curve AB is 9 units.
How long is curve RS?

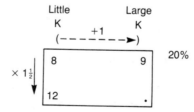

Here the multiplier (1½) is no more complex than in Items II and II' (though the answer, 13½, is slightly more awkward). However, the item proved to be far more difficult, and a substantial number of children gave the answer 13 (40 per cent) by using what Piaget (1967) and Karplus (1974) have called the 'Addition Strategy' (8 + 1 = 9→12 + 1 = 13, or 8 + 4 = 12 → 9 + 4 = 13). 40 per cent of children also used

this strategy in Item III, and an even greater proportion
(51 per cent) used it in Item V, (Karplus's 'Mr Short and
Mr Tall').

As a final consideration, we might look again at the
addition strategy. The question devised by Karplus, and
already referred to, is:

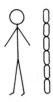

You can see the height of Mr Short
measured with paper clips.

Mr Short has a friend Mr Tall. When we
measure their heights with matchsticks,
Mr Tall's height is six matchsticks. How
many paper clips are needed for Mr Tall's
height?

('Correct' ratio answer is 9, addition strategy answer is 8).

Karplus (1974) commments that the addition strategy answer is
frequently given. Hart (1978) states that about 50 per cent
of children used this incorrect strategy not just for this
task, but consistently on three or four similar items.
Moreover, these were not the least able children: they were
achieving at the 50-60 per cent level on items with 50 per
cent over-all facility. The consistent, incorrect method was
satisfactory to them.

In conclusion, it might be said that the understanding
of numerical relationships involving ratio develops very
slowly. Certain types of task, involving only doubling or
halving, are relatively easy, and are not indicative of an
ability to answer other ratio questions. Presenting children
with harder ratio questions seems to drive them to search for
some method - any simple method - of producing numbers of
about the right size, and the addition strategy often appears.
Generally speaking, if a task requires fractions other than
halves, it becomes much more difficult: particular operations
on such fractions may not be recognized as appropriate in the
context; and the techniques, if attempted, may be remembered
incorrectly.

Directed number

In view of the known considerable difficulty of this area and
its fundamental importance to the understanding of algebra
and of the graphical representation of many functions and
quantities, such as speed, distance and time, it is surprising that so little research exists. Some work which bears on
it, though not explicitly, is reported in subsection 3 above,
where the tendency to commute the subtraction in an item
such as 8 - 13 + 20 is widespread until the later years of
the secondary school.

The two APU surveys and the CSMS results provide somewhat less information in the area of negative numbers than in
other fields discussed earlier. The relevant CSMS test was
not fully developed, although some results have been published and discussed by Küchemann (1980b).

A very common representation of directed numbers is as
points on the number line, and the CSMS results suggest that
children are able to cope with this at secondary level: the
items shown below were answered correctly by 95 per cent
and 94 per cent of 14-year-olds respectively.

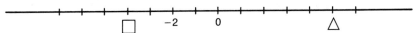

What number would go in △? . . . .

What number would go in □? . . . .

The facilities quoted by the APU surveys are not so encouraging: they are for the slightly different task of actually
placing a number at the appropriate point on the line.   80
per cent of 15-year-old children were able to locate a
positive integer and zero, while a negative integer was
correctly placed by only 65 per cent. Corresponding figures
for similar items in the Primary Survey were just under
65 per cent for locating a positive integer, about 50 per
cent for zero and under 40 per cent for a negative integer.

Further work on directed numbers is concerned with the
understanding of operations on these numbers. In the CSMS
test, various approaches or models are provided to explain
these operations, and some items explicitly mention moves or
shifts, for example 'two steps forwards and then three steps
backwards is the same as the one move ...'. The APU Secondary Survey includes six items expressed simply as computations with integers, so that the results are difficult to
compare. In both cases however, when negative integers are
involved, addition is considerably easier than subtraction:
in the CSMS work, questions involving addition have a
facility of 80 - 90 per cent, while subtractions vary from
36 - 77 per cent; while the APU survey quotes 65 per cent for
adding negative integers, and about 50 per cent for subtracting a negative integer. Both studies show that a common
incorrect approach to subtraction is to start by subtracting
the numbers followed by an attempt to determine the sign of
the answer by some rule. Thus, in the APU survey, about half
correctly answered 7 - (⁻3) while over 20 per cent gave the
answer as 4.   In the CSMS test the item ⁻6 - ⁻3 had a
facility of only 36 per cent, the answers ⁺3 or ⁻3 being
given by 47 per cent of 14-year-old children. Facilities for
multiplication, however, are somewhat higher: around 65 per
cent in the APU survey and 80 per cent in CSMS. These are
close to the facilities for addition, even though the latter

operation is much easier to model. It seems likely that children are able to use rules for combining signs in multiplication tasks without the confusion which such rules cause in subtraction items.

The ability of 11-year-old children to locate negative numbers on a number line appears relatively high considering that many teachers did not consider the questions on negative numbers to be appropriate for their pupils. In the APU Primary Survey, 47 per cent consider the questions 'hardly appropriate' or 'not appropriate at all', and a further 41 per cent appropriate 'only to a limited extent'.

Moreover, an understanding of the concepts involved may be present in familiar situations, without the explicit use of negative numbers. Children's success with items of this sort has been investigated by Lunzer, Bell and Shiu (1976), who include the following item:

> Christine was saving to buy her Gran a birthday present. She had saved 15 pence. The present she wanted to buy cost 27 pence. Luckily, Christine's uncle gave her 20 pence.

> How much money would Christine have after she had bought her Gran's present?

> Write a number sentence to show how you got your answer.

The numbers of correct responses were:

Age	11/12	12/13	13/14	14/15 (lower set)	14/15 (upper set)	Total
No. of children	25	25	28	26	27	131
Solution of embodied problem	19	19	20	17	24	99
Composition of number sentence	15	20	17	19	26	97

In spite of the order in which the numbers are given, it is fairly easy for most children to see that the 15 and 20 can be 'collected and added' and the 27 then subtracted. The expressions 15 - 27 + 20 and 15 + 20 - 27 are equivalent in this case. The 'story' can be understood as a whole, whereas numerical expressions may have to be worked through in sequence. It is somewhat more difficult to understand that in 2 - 5 + $^{+}$11, the 2 and 11 can be combined before subtracting the 5; in fact only 28 per cent of the same pupils were able to rearrange this correctly.

It is perhaps the lack of a good familiar model for subtraction which makes this operation on directed numbers relatively difficult. Addition is easily modelled and can be performed with an intuitive understanding. Multiplication is less easily explained, but rules can be provided which are fairly foolproof. In subtraction, however, models are complicated and rules are very likely to be confused and misused. Many secondary school children are not going to be able to cope with subtraction in an abstract way; and at the same time they are not provided with an accessible model.

Other research on the understanding of number

Saad and Storer (1960) made a broad survey of the under-standing of the ideas and skills of the grammar school mathematics curriculum as it was in 1956. In some respects these results supplement those currently available from more recent research, but more than this, they have some compara-tive interest. There are few questions which can be mapped with any degree of exactness onto currently tested items, but as one example, questions on the addition and subtraction of directed numbers using the number line show lower percentages of success for this grammar school sample than appear in the current APU Secondary Survey covering the whole ability range.

The work of Collis extends from number into algebra, and is mentioned below in the section on algebra. However, there are some results on, for example, the development of the recognition of a sum, such as 56 + 21, as representing not simply an instruction to perform a calculation but also as representing a number which results from this; and similarly for 56 ÷ 37, and other numbers combined by simple operations. This appears in Collis's book, *A Study of Concrete and Formal Operations in School Mathematics* (1975a).

Relevant articles in recent issues of the *Journal for Research in Mathematics Education,* which are not otherwise mentioned here, include one by Wheatley (1976) which shows that, in performing additions of figures arranged in vertical columns, the method of working directly up or down the column is in general substantially more efficient, as measured by the number of correct answers, than the method of picking out digits which together make tens. However, the latter is the method generally preferred and which most people tend to adopt if not constrained.

Some other relevant material is contained in Steffe (1975).

3. ALGEBRA

Under this heading we shall consider both the algebra of numbers and that of structure, and a subsequent heading will consider functions and graphical representation. However, apart from the small section in the APU survey, which we report below, and a more substantial CSMS test on vectors and on matrices, there seems to be no research on the understanding of matrices or structural algebra.

We begin the discussion of the understanding of the algebra of numbers by showing the cluster of items on equations from the APU Primary Survey, and on substitution from the Secondary Survey:

Items from generalised arithmetic
Item cluster: equations
Mean sub-category score—38 per cent

Response analysis		Item facility	
Incorrect	6%	88%	M1 Find which number ☐ stands for.
Omitted	6%		$12 - \square = 8$ $\square = \ldots\ldots\ldots$
Incorrect	19%	75%	M2 Find which number △ stands for.
Omitted	6%		$51 + \triangle = 90$ $\triangle = \ldots\ldots\ldots$
Incorrect	30%	63%	M3 n stands for a number.
Omitted	7%		$n + 4 = 21$
			so $n + 5 = \ldots\ldots\ldots$
Incorrect	39%	51%	M4 B stands for a number.
Omitted	10%		$B - 9 = 21$
			so $B - 10 = \ldots\ldots\ldots$
Incorrect	50%	19%	M5 Fill in the values of M in this table according to the equation $M + N = 4$
Omitted	31%		

N	0	1	2	3	4
M	4				

Items from traditional algebra
Item cluster: substitution
Mean sub-category score 35 per cent.

		Item facility	Response analysis	
$x = a + b - c$ Find x if $a = 1$, $b = 7$ and $c = 3$ $x = 5$	U1	77%	Incorrect: 11 Other Omitted	4% 13% 6%
If $a^x = 4$ and $a^y = 5$ what is the value of $a^x + a^y$? 9	U2	64%	Incorrect: a^9 Other Omitted	8% 16% 12%
$y = d^3$ Find y if $d = 3$ $y = 27$	U3	38%	Incorrect: 9 Other Omitted	19% 27% 16%
$x = a + b$ $y = a - b$ $z = 2(a + b)^2 - (a + b)(a - b) + 3(a - b)^2$ Write z in terms of x and y $z = 2x^2 - xy + 3y^2$	U4	30%	Incorrect Omitted	31% 39%
If $a = 3$, $b = -2$ and $c = 7$, evaluate $3b^2 - abc$ 54	U5	13%	Incorrect: -30 -54 30 Other Omitted	4% 1% 1% 63% 18%

In considering the understanding of the algebra of numbers, we are concerned with the development of the ability to interpret and handle letters and other symbols, which may stand for objects, unknown (but definite numbers), 'generalized' numbers or variables, and at a later stage operations, relations or abstract entities constructed by defined relations. In this topic, it is useful to recognize which types of interpretation and operation are more difficult than others, to understand the ways in which children interpret or misinterpret the symbols at different stages of development, and to identify the particular forms of interpretation and procedure which appear likely to produce the desired aspects of competence in dealing with algebraic tasks.

Collis (1975b) identifies various ways in which children interpret letters in generalized arithmetic, refining the terms 'variable' and 'unknown', and developing a more precise classification of these interpretations. His ideas were used as a basis for constructing the CSMS algebra test. The following six categories for describing different uses of letters in that test are given by Küchemann (1978):

1.   Letter evaluated.
2.   Letter ignored (letter not used).
3.   Letter as object.
4.   Letter as specific unknown.
5.   Letter as generalized number.
6.   Letter as variable.

The items on the test are ordered into four groups representing levels of difficulty, which are only approximately related to the six categories described. It may be noted that categories 4, 5 and 6 can be seen to represent different theoretical uses of letters in algebra, but categories 1, 2 and 3 are something different. They indicate ways in which children may interpret letters to avoid the formal theoretical understanding implicit in the topic. Thus, items which can be handled successfully by the interpretations and strategies of categories 1, 2 and 3 prove accessible to a significantly larger proportion of secondary school pupils than those which demand an appreciation of the later categories. In some situations, the prevalent wrong answers are indicative of the kind of interpretation applied: thus, in the following item, children are disposed to think of the letters as representing objects (i.e., sides of the figure):

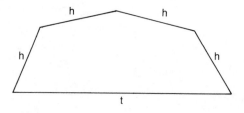

Perimeter = ...

(68% correct, 20% 4ht or hhhht.)

134

Piagetian Sub-Stage	Item	Question Number on test	% Correct	EVALUATED	IGNORED	OBJECT	SPECIFIC UNK	GNRLSD NMBR	VARIABLE	QUESTION	Common Wrong Answers	%
Early Concrete	A	5i	97	x						If $a + b = 43$, $a + b + 2 = $		
	B	9i	94		x					$p = $		
	C	6i	92	x						If $a + 5 = 8$, $a = $		
	D	7i	91							$A = $		
	E	7ii	89							$A = $		
	F	1i	88							If $x \to x + 2$, $6 \to $		
Late Concrete	G	5ii	74	x						If $n - 246 = 762$, $n - 247 = $	763	13
	H	2	72	x						smallest, largest of: $n + 1$, $n + 4$, $n - 3$, n, $n - 7$		
	I	4i	68	x						Add 4 onto $n + 5$	9	20
	J	7iii	68			x				$A = $		
	K	9ii	68			x				$p = $	$p = 4ht$ or$hhhht$	20
	L	9iii	64			x				$p = $	$p = 2u16$ or $uu556$	16
	M	11ii	62	x						If $m = 3n + 1$ and $n = 4$, $m = $		
	N	11i	61	x						If $u = v + 3$ and $v = 1$, $u = $	2	14
Early Formal	O	5iii	41				x			If $e + f = 8$, $e + f + g = $	12	26
	P	14	41				x			If $r = s + t$ and $r + s + t = 30$, $r = $	10	21
	Q	9iv	38				x			n-sided polygon, each side of length 2; $p = $	36, 38, etc.	18
	R	4ii	36				x			Add 4 onto $3n$	$7n$	31
	S	16	30					x		What can you say about c if $c + d = 10$ and c is less than d	4 only	39
	T	18ii	25					x		Is $L + M + N = L + P + N$ always, sometimes or never true?	never	51
	U	20	22				x			Cakes cost c pence each and buns cost b pence each. If I buy 4 cakes and 3 buns, what does $4c + 3b$ stand for?	4 cakes and 3 buns	39
Late Formal	V	4iii	17				x			Multiply $n + 5$ by 4.	$n + 20$	31
	W	7iv	12				x			$A = $	$e + 10$, $10e$, $7e$	28
	X	22	11				x			"blue pencils and red pencils" (see text).	$b + r = 90$	17
	Y	17i	5				x			Mary's basic wage is £20 per week. She is also paid another £2 for each hour of overtime that she works. If h stands for the number of hours of overtime that she works, and if W stands for her *total* wage (in £'s) write down an equation connecting W and h:..................	$W + h$ or $W = 20 + h$	27
	Z	3	6						x	Which is larger, $2n$ or $n + 2$? Explain.	$2n$	71

The item 'a + 5 = 8, a = ?' is correctly performed by 92 per cent of third-year secondary school children. However, an investigation by Lunzer, Bell and Shiu (1976) involving problems of this kind suggests that younger children have a limited understanding of the equals sign. Many seven to eight-year-old children do not interpret '=' as 'is the same as' but rather as 'makes' or 'gives the answer .....'. Thus, children who can solve 4 + □ = 7 may not be able, at a certain stage, to make any sense of 7 = 4 + □ . The greater difficulty of the second form persists with some pupils into the secondary school years.

Shiu (1978) has analysed the approaches of children to items where there is a choice between ignoring and evaluating an unknown. For the item,

$$E + 17 = 36$$
$$\text{so } E + 12 = \square,$$

some subjects clearly evaluate E from the first equation and substitute it in the second; others reason thus: 'Well, 12 is 5 less than 17 so the number in the box must be 5 less than 36.' Shiu calls the first strategy 'sequential closure', and the second a 'displacement strategy': the latter demands a greater degree of Collis' 'acceptance of lack of closure'. In general, subjects at the age of 11 - 12 tended to favour sequential closure, while mathematically competent subjects of 16 - 17 were more likely to adopt the displacement strategy, although many still preferred sequential closure.

The use of letters as objects is a very effective way of reducing the difficulty of certain algebra problems. However, the continuing tendency to regard letters as symbols for objects rather than numbers appears to be a significant stumbling block in learning algebra. Thus the CSMS test includes the item:

'Blue pencils cost 5 pence each and red pencils cost 6 pence each. I buy some blue and some red pencils and altogether it costs me 90 pence.

If b is the number of blue pencils bought, and if r is the number of red pencils bought, what can you write down about b and r?'

This has a facility of only 11 per cent; the most common answer is $b + r = 90$, which might be 'translated' as 'blue pencils plus red pencils cost 90p' rather than understood as a relationship between numbers.

In some problems the English sentence and the mathematical statement are closely matched, and such direct translation may give the correct equation, but as Galvin and Bell (1977) explain, this may not imply that the equation is

136

correctly interpreted. Thus, the following translation from English to a symbolic shorthand expresses the correct relationship but is not understood in the correct terms:

(1 red pencil) (and) (6 blue pencils) (cost) (23p)

$1r$ + $6b$ =   23

A number of the difficulties involved in 'specific unknown' problems can be explained in this way. Similarly in 'generalized number' examples, many children clearly want to attach a single value to each letter: this shows clearly in responses to the item:

'What can you say about c if $c + d = 10$ and $c < d$?'

The item which Küchemann considers best tests the use of a letter as a variable is: 'Which is larger, $2n$ or $n + 2$?' It has a facility of 6 per cent. 71 per cent wrote that $2n$ was larger for a reason like 'because it is multiply'. Others made an inference from a single value. Rather than describing a single relationship here, it is necessary to explain how one relationship depends on another: the answer depends on whether n is greater or less than 2. Relationships of this kind are termed 'second-order' by Inhelder and Piaget (1958).

The analysis of the Algebra test relates the work to the Piagetian distinction between concrete and formal operational thought. Ingle and Shayer (1971) and Malpas and Brown (1974) have suggested that much of secondary teaching assumes the possibility of formal operational thinking of the kind described here. Yet in this algebra test, the percentages of 2nd, 3rd and 4th year pupils able to cope consistently with at least the early formal items were about 15 per cent, 35 per cent and 40 per cent respectively. With this in mind, the results could provide some kind of framework for teachers to interpret their pupils' efforts.

There are a number of studies which identify levels of difficulty in tasks which involve interpreting, reorganizing or solving equations. The following item was used by Firth (1975) with two classes of secondary pupils, one a fourth year of ability around 'O'-level/CSE borderline, the other a third year, somewhat more able. The percentages given here are those from the fourth year but those for the third were very similar. Nevertheless, we cannot attach much significance to small differences here.

$$P = R + S - T$$

P − R =		50
P + T =		76
P − S + T =		65
S − T =		29
T − S =		35

The most striking result here is the large drop in facility when it is required to produce an answer in which P does not figure on the left-hand side. A possible explanation might be that even amongst this age group the equals sign is not fully seen as reversible. The other sharp difference in these results is between the first and second items, which shows that transferring R to the left-hand side is harder than transferring T. This may be because T has a sign which can be changed on transfer whereas R does not obviously have a sign. Alternatively, if the mental explanation is that we add T to both sides, this is easier to do at the end of the expression than it is to do for R, since subtracting R from the right involves questions of associativity. Whether or not these explanations apply, the clear differences between the scores certainly show that we are dealing with something other than simply knowing or not knowing a simple manipulative rule. There are factors affecting the understanding and use of these rules which often go unrecognized and therefore untreated in the course of teaching.

Another difficulty, perhaps more generally recognized, appeared in another item in the same test. This said 'x is any number; write the number which is 3 more than x'. In this case about a third of all pupils studied could only respond by choosing a number for x and using this in the subsequent sections. Interviews supported the conclusion that the pupils were effectively saying 'I can't do this until you tell me what x is'. They did not realize that the purpose of algebraic expression is to provide a means of recording what is known about a number without knowing what the number is. In other words, these children were unwilling to operate on a specific unknown, which one might say requires, in a certain sense, the ability to tolerate an uncertainty.

The following items come from work by Collis (1975b):

Decide whether the following statements are true always,
sometimes or never. Put a circle round the right answer.
If you put a circle round 'sometimes' explain when this
statement is true.

AGE	10	11	12	13	14	15
	100	80	90	100	100	100

1. $a + b = b + a$

always
never
sometimes, that is when ....
. . . .

	70	60	70	70	100	100

2. $a + b + c = c + a + b$

always
never
sometimes, that is when ....
. . . .

	0	10	0	10	33	50

3. $a + b + c = a + b + d$

always
never
sometimes, that is when ....
. . . .

	0	10	10	20	33	50

3(a) $m + n + q = m + p + q$

always
never
sometimes, that is when ....
. . . .

	0	10	10	20	33	50

4. $a + 2b = 2a + b$

always
never
sometimes, that is when ....
. . . .

	0	0	0	20	67	40

5. $a + 2b + 2c =$
    $a + 2b + 4c$

always
never
sometimes, that is when ....
. . . .

	0	0	0	10	0	0

These tests were given to 30 pupils of each age from 10 - 15.
Interviews were conducted with ten pupils from each age group,
except for the 14-year-olds from whom the interview results
are from six people only, and it is these children's results
that are reproduced above. The first observation might be
that even the ten-year-olds have no difficulty in understanding the meaning of the question or of the symbols. It is not
necessary to have been taught algebra in a formal sense to be
able to interpret questions such as this. They can accept
that a letter may be used to stand for a number and almost all

139

will agree that $a + b = b + a$, though some of these are more doubtful about whether the equality will still hold if the order of the numbers is changed from $a + b + c$ to $c + a + b$. The second observation concerns the dramatic increase in difficulty seen in item 3. The dominant response here was 'never', the assumption being that c and d being different letters could not be the same number. To respond correctly to this item demands the concept of both c and d running through the whole set of possible numbers and also the willingness to tolerate the ambiguous answer 'sometimes'. The previous items could be answered successfully, by giving specific values to a, b and c. The concept of c and d actually standing for a variety of possibilities is clearly very considerably more difficult. The third observation here concerns the even greater difficulty of item 5 which was answered correctly by only one pupil out of the entire 56 interviewed. Here the problem is that of remembering zero as a possible value to attribute to c. This is regarded as a special number and often forgotten in such cases.

Some further studies of Collis (1978) are concerned with elementary abstract mathematical systems. In one such study, an operation * is defined by

$$a \; * \; b \; = \; a \; + \; (2 \times b).$$

In test 1 below, the children were asked to indicate when the statement would be true; in tests 2 and 3 they were asked to mark the statement true or false.

Item	Test 1	Test 2	Test 3
1	$a * b = b * a$	$4 * 6 = 6 * 4$	$4728 * 8976 = 8976 * 4728$
2	$a * (b * c)$	$5 * (4 * 6)$	$982 * (475 * 638)$
	$= (a * b) * c$	$= (5 * 4) * 6$	$= (982 * 475) * 638$
3	$a * x = a$	$4 * 5 = 4$	$4932 * 8742 = 4932$
4	$a * (b + c)$	$3 * (4 + 6)$	$6836 * (935 + 2397)$
	$= (a * b) * c$	$= (3 * 4) * 6$	$= (6836 * 935) * 2397$
5	$a + (b * c)$	$3 + (4 * 6)$	$572 + (865 * 749)$
	$= (b * c) + a$	$= (4 * 6) + 3$	$= (865 * 749) + 572$

The responses reveal that a common approach to test 1 up to about 16 years of age is to ignore the defined operation and try out ordinary arithmetical operations, providing answers such as 'true when * means +'. Two other incorrect categories of response can be identified. One is the persistent 'can't tell' category. The other characterizes a group of children who recognize the need to use the operation as defined but have not sufficient control to handle the system. The responses of three subjects in this category are given, for test 1.

Example 1
Item

1	This will be true when $a = b$
2	True when a and c are equal
3	True when $A = 1$ or 0
4	True when $(b + c) = (a * b)$ and $c = a$
5	$(b * c) - a = a - (b * c)$

Note Correct in item 1; an interesting confusion in items 2, 3, and 4; the response to item 5 appears to be an attempt to work it out.

Example 2
Item

1	When $b * a = a + 2 \times b$
2	When $a * b = b * c$
3	When $a = a + 2x$
4	True if $a + 2b = b + 2c$, that is $a * b = b * c$
5	True if $a * b = b * c$

Note Interesting confusion with same idea recurring.

Example 3
Item

1	$a + 2 \times b = b + 2 \times a$
2	$a + 2(b + 2 \times c) = (a + 2 \times b)2 \times c$
3	$- -$
4	$a + 2(b + c) = (a + 2 \times b) \times 2 \times c$
5	$a + (b + 2 \times x) = (b + 2 \times c) + a$

Note The student could work thus, but seemed unable or unwilling to draw a conclusion in any example.

Collis reports the proportion of subjects in each response category by age group as follows:

Proportion of subjects in response categories by age group

Age	Can't tell (1)	+ or – substitution (2)	Attempt to work with system defined (3)	Worked correctly with system defined (4)
17	0.00	0.14	0.23	0.63
16	0.06	0.52	0.16	0.26
15	0.03	0.68	0.29	0.00
14	0.09	0.80	0.11	0.00
13	0.10	0.87	0.03	0.00
12	0.03	0.94	0.00	0.03
11	0.12	0.88	0.00	0.00
10	0.17	0.83	0.00	0.00
9	0.00	1.00	0.00	0.00
8	0.27	0.73	0.00	0.00
7	1.00	0.00	0.00	0.00

The judgment as to which category should be allotted was made on the basis of three correct responses out of five in test 1. By nine years of age, children generally substituted a familiar system for the defined one; and, by 16, 42 per

141

cent attempted to work with the given system, 26 per cent
successfully. A noticeable strategy of the 16-17-year-olds
is their use of general solutions from test 1 items to give
correct responses in the other tests. In contrast, the
younger children persisted in regarding the tests as independent, even when prompted by the investigator.

Some research studies have been published which consider
the appropriateness of different frameworks in which algebraic equations may be manipulated and solved. Adi (1978)
discusses the relative merits of two different methods, and
also lists a number of studies which compare procedures -
she gives the following example:

Solve:
$$14 - \frac{15}{7 - x} = 9$$

(1) *Cover-up or reversal method:*

14 minus what equals 9?	(5)
15 divided by what equals 5?	(3)
7 minus what equals 3?	(4)

Solution: 4

(2) *Formal method:*
Multiply both members of the given equation by
(7 - x)

$$14(7 - x) - 15 = 9(7 - x)$$
$$98 - 14x - 15 = 63 - 9x$$
$$83 - 14x = 63 - 9x$$

Add (14x) to both members,
$$83 = 63 + 5x$$
Subtract (63) from both members,
$$20 = 5x$$
Divide both members by 5,
$$4 = x$$

Possible solution: 4
Check whether 4 is a solution of the given equation

$$14 - \frac{15}{7 - (4)} = 9$$

It is difficult to point to any generally useful conclusions
from such studies: they are neither extensive nor convincing.
Moreover, it is inappropriate to rely only on the measurement of children's skill by an immediate post-test; as in
many other fields, the development of meaningful and structured understanding, carrying the expectation of a high
degree of retention, is probably more important.

An investigation of the mathematical achievement of 16-
year-old students in Sweden is reported by Ekenstam and
Nilsson (1979); the results quoted are for students going
on to further academic education, and so reflect the capab-
ilities of roughly the top 20 per cent of the population in

academic terms.   This study provides a substantial amount
of material in various subject areas, with facilities quoted
for questions in much the same way as the CSMS and APU results
are given.   The work on manipulative algebra is of particular
interest in that, by careful mixing of questions and allocation of tasks to different samples of children, the investigations have been able to quote facilities for solving
equations each of which can be considered as a stage in the
solution of one original equation.   Thus questions can be
considered such as: How many students master the last step
and no others?   All steps except the original equation?   Do
more pupils give the correct answer if $7x = 6$ is changed to
$7x = 14$ to give a whole number solution?   Would changing
x to t affect the difficulty?   An example is shown below:

Test item	Comment	Facility (%)
(1) $\dfrac{3x-2}{2} = \dfrac{x}{3}$	Original equation	28
(2) $3(3x - 2) = 2x$	The first step performed	70
(3) $9x - 6 = 2x$	The second one performed	74
(4) $9R - 6 = 2R$	As (3) but x is changed to R	64
(5) $7x - 6 = 0$	A possible step after (3)	71
(6) $7t - 6 = 0$	As (5) but x is changed to t	72
(7) $7x = 0$	Final step	77
(8) $243x = 242$	As (7) but the coefficients are changed	69

Some of the results show a lack of awareness of the most
basic notions, e.g., that a letter stands for a number.
Thus, while almost every student could have written 15/15
in its lowest terms, only about 50 per cent could simplify
a/a.   The facilities for a/a^2 and a^2/a are 36 per cent and
87 per cent respectively: in the first one the numerator
'disappears' and a common incorrect answer is 0.   It is
possible that a/a is correctly simplified, but with an incorrect intermediate step of reasoning: $a^2/a = a/0 = a$.

Another notable feature of the results is the effect on
the facility of the nature of the solution.   Thus $4/x = 3$
(48 per cent) proves harder than $30/x = 6$ (82 per cent) or
even $\dfrac{14}{x+2} = 2$ (58 per cent): the latter two examples have
whole number solutions.

The Journal of Children's Mathematical Behaviour
includes some examples and analyses of pupil' strategies in
solving algebra problems.   Davis, Jockusch and McKnight
(1978), for example, give detailed transcripts of interviews
and conversations; these show up individual strategies and

misconceptions, such as the 'cancellation'

$$\frac{3x}{x} = 2x$$

Also a number of obviously useful strategies which are not frequently used are identified, such as 'checking with numbers' which could presumably reduce the prevalence of errors such as

$$(A + B)^2 = A^2 + B^2.$$

The discussion, however, has little coherent structure: perhaps the most obvious conclusion is that many errors are caused by children who have developed skill in manipulating meaningless symbols being disinclined to think in terms of meaning or to consider that the symbols represent numbers.

Current work at Nottingham by Bell and others is experimenting with an algebra course in which the concepts and relations underlying the usual rules for transforming equations form the framework of the course, and the emphasis is on a substantial amount of the learning of meanings of expressions before concluding with the development of fluency in the use of rules.

4. FUNCTIONS AND GRAPHS

The APU Primary Survey reports the following set of items concerning graphical representation:

Response analysis		Item facility	*Items from probability and data representation* *Item cluster: graphs and bar charts* *Mean sub-category score–48 per cent.*
a) Incorrect	8%	90%	L1
Omitted	2%		
b) Incorrect	3%	95%	
Omitted	2%		
c) Incorrect	51%	45%	5 girls made a chart to compare their weights when they were in Class 6 and again a year later when they were in Class 7.
Omitted	4%		a) Who was the *lightest* girl in *Class 7*? b) Who was the *heaviest* girl in *Class 6*? c) Who *gained* most weight during this year?

144

a) Incorrect	36%	50%	
Omitted	14%		
b) Incorrect	57%	23%	
Omitted	20%		
b) Incorrect	64%	23%	
Omitted	18%		

L2

This graph shows the new prices compared with the old prices.
a) What is the new price of something with an old price of £1.50? ....
b) What was the old price of something with a new price of £1.20? ....
c) What is the old price of something with a new price of £1? ....

A wider range of comprehension and interpretation of graphs representing complex situations at the secondary level has been undertaken by Janvier (1978). Some of this is reported in a paper by Bell (1979), and full details are contained in Janvier's thesis, which is available in duplicated form. Work by Orton and by Thomas on the understanding of some of the more technical aspects of function is reported in Lovell's chapter in Rosskopf (Ed): *Piagetian Cognitive Development Research and Mathematical Education*, and also in *Six Piagetian Studies* (Rosskopf, 1975). There is a difficult book by Piaget *et al.* on the epistomology and psychology of functions which is concerned with the development of the understanding of functions connecting numbers and quantities through the ages of 5-12. The understanding of limits and concepts of derivative and integral have also been studied by Orton; this material is as yet unpublished, but useful material is available from Orton at Leeds.

5.   PROBABILITY AND STATISTICS

This is another area which we have been unable to deal with in detail. There exist books on the subject by Piaget and Inhelder (1951) and by Fischbein (1975). There is an article (Wood and Brown) based on the results in particular items in an O-level examination. There is also a current SSRC project at Loughborough directed by Green, which aims to produce an analysis of understanding in this area.

Although the representation of data in graphical form features extensively in mathematical work with young children, less attention appears to have been given to developing children's ability to appreciate the notion of probability

145

Items from probability Mean sub-category score–40 per cent.		Item facility	Response analysis	
In a very large sample of women who have exactly two children, the probability of having a girl is the same as the probability of having a boy. What is the probability that a mother in this group has one girl and one boy? A. $\frac{1}{4}$ B. $\frac{1}{3}$ C C. $\frac{1}{2}$ .... D. $\frac{3}{4}$	W1	70%	A B C D Other Omitted	12% 9% 70% 4% 1% 4%
A ball is drawn at random from a box containing 5 red balls, 2 white balls and 6 green balls. Each ball is equally likely to be chosen. What is the probability that the ball is green? A. $\frac{6}{7}$ B. $\frac{6}{13}$ B C. $\frac{7}{13}$ .... D. $\frac{13}{7}$	W2	65%	A B C D Other Omitted	20% 65% 7% 4% 1% 3%
A spinner is equally likely to point to any one of the numbers 1, 2, 3, 4, 5, 6, 7. What is the probability of scoring a number exactly divisible by 3? $\frac{2}{7}$	W3	48%	Correct: $\frac{2}{7}$ 2:7 2 in 7 Incorrect Omitted	42% 2% 4% 40% 12%
What is the probability of drawing an ace at random from a complete pack of shuffled cards, turned face downwards. $\frac{1}{13}$	W4	47%	Correct: $\frac{1}{13}$ $\frac{4}{52}$ Incorrect Omitted	24% 23% 41% 12%
A card is drawn at random from a standard pack of 52 playing cards. What is the probability of drawing a red card or an ace? A. $\frac{2}{52}$ B. $\frac{26}{52}$ C C. $\frac{28}{52}$ .... D. $\frac{30}{52}$	W5	30%	A B C D Other Omitted	19% 28% 30% 17% 1% 5%

and compare likelihoods. The APU Primary Survey contains just three items of this nature: in one item about 35 per cent expected to guess the outcome when a die was rolled more often than when a coin was tossed. A page from the Secondary Survey is shown (see page 146).

Recent research by Falk, Falk and Levin (1980) has examined children's ability to select the greater probability in a game of chance, and found that at about the age of six children can begin to select the greater of two probabilities systematically. The understanding of the chance concept tested verbally, however, lagged significantly behind efficient performance in the choice tasks. Some questions were asked to discover whether children expected greater success (in a game of chance):
1. if they practised for a fortnight;
2. when the game was played by a child three years older;
3. if a roulette dial (or top) were spun with eyes closed.
Some responses are quoted:

' Ilana (4:9): "With open (eyes) it is easier, you win better because then you see where it reaches" Ilana answered the question of how a girl older than herself might succeed in the same lottery, as follows: "I think she can close her eyes because she is already big, so it comes out well for her. I don't understand the game too well, so I should do it with my eyes open."
Tami (5:3): "A big girl would win better than me, because she is older, because she was playing all the time since she was small." This is a typical response expressing the opinion that one can improve with a lot of experience.
Shlomo (5:5): "A big boy would win more prizes because he is *stronger*."
Naomi (6.3): "You can win better with open eyes, because it's easier," and "A big girl will surely win more, she is smarter, she has got more brains."
Here is a girl who scored positively on the "closed eyes" question and negatively on what would happen in a fortnight:
Raya (7:8): "With closed eyes you win just the same, because it does not help to open your eyes, you can't make it succeed with open eyes either" and "In two weeks I will probably win more because I shall *know better* to turn the dial so that the pointer will come to the right place."
Finally, here are two correct responses: Rachel (6:9): "With closed eyes ... just the same, because the top is doing it and not me." Ari (10:5): "in two weeks I shall win just the same, because it works only on luck!"'

Fischbein (1975) claims that some intuition of chance is present even before the age of six years; but it appears that this sometimes deteriorates during primary school years. Fischbein attributes this to the authoritative atmosphere of school, in which children may be encouraged to look for deterministic answers. It may be that the balance needs to

be restored in favour of indeterminism: Falk *et al.* suggest
that replies like 'unlikely' and 'maybe' should be part of the
repertoire for certain questions, as for example:

'Will the telephone ring in the next hour?'
'Will the plane land on time?'

These are situations characterized by several possible outcomes, for which only degrees of belief can be given.

6.   SPACE AND GEOMETRY

Two books (Martin, 1975a, and Lesh, 1978), published by the
ERIC centre in Ohio, provide a summary of some recent
research on the development of spatial and geometric ideas.
Some articles from these are quoted here; others are more
concerned with proof in geometry and are considered in the
next chapter.

There is also a review article on spatial abilities and
mathematics education by Bishop (1980) which identifies a
number of specific issues. Of particular interest is the
discussion of teaching experiments, some of which are said
to have achieved considerable success in improving spatial
ability.

In the following discussion, various ways in which
children's appreciation of shape and space develops, are
considered. In particular, three aspects of this development are identified:
1.   the change from spatial 'egocentrism' to the ability to
     take another point of view;
2.   the recognition of certain topological properties prior
     to projective or Euclidean (metric) properties;
3.   the transition from global recognition to analysis of
     relationships and properties.
The first two of these are described by Piaget and Inhelder
(1956) and have attracted a lot of research interest: the
third, while often not made explicit, is clearly identifiable
in a number of studies. In the present section, all three
forms of development are considered in the context of children's ability to recognize, draw or reconstruct a geometrical
configuration; but it may be noted that they may also have
relevance (and particularly (3)) to children's view of what
constitutes 'justification' in geometry and to their appreciation of the need for proof. These latter aspects are
discussed in Chapter Eight.

The basic task used by Piaget and Inhelder to study
spatial egocentrism involves showing children a 3-dimensional
model of three mountains of different size and colour, and
then asking them to select from ten pictures the view seen
by a doll placed at different locations around the model.
Piaget and Inhelder classify children's responses into three

age-related stages. Between about four to seven years, children respond egocentrically by choosing their own view as that of the doll. Then there is a transitional stage around seven to eight years at which children begin to recognize that the doll's perspective is different from their own but are unable to handle accurately the spatial relationships involved. Then, at about nine or ten years of age, their choice of picture is nearly always correct. These conclusions are based essentially on this one task, and give support to the claim that spatial egocentrism is a feature of young children's thinking: the results have been thoroughly replicated (see, for example, Laurendeau and Pinard, 1970).

A number of studies provide evidence that changing certain features of this task can significantly reduce the proportion of egocentric responses, even amongst groups of very young children. Borke (1975) asked children to revolve a duplicate display to match the doll's viewpoint, and also compared children's ability to handle the mountain model with their ability to cope with 'naturalistic' groups of toys (including ducks on a lake and cows pulling a wagon). The results of this study show that the mountain task is significantly more difficult than the others. It would appear that the use of familiar, animate, asymmetrical objects makes the task easier; and the fact that young children achieved some success with the mountains suggests that actually turning a display is easier than choosing a two-dimensional representation.

Other investigations by Cox (1975) and Fehr (1979) show that replacing the doll by a human observer makes the task easier. The view put forward by Donaldson (1978) is that egocentric responses depend not only on the conceptual complexity of a task, but on whether it makes 'human sense'. Her task, in which children are asked to show where it would be possible to hide from the view of two policemen, can be seen in these terms, although it does seem intrinsically easier than the mountains problem.

It does appear that the tendency to respond egocentrically is at least partly dependent on age and does become less in older children; but, like most other phenomena, it is also affected by the complexity and context of the particular tasks. Thus children are less likely to make egocentric errors in natural, real and familiar situations.

Another notion attributable to Piaget and Inhelder is that the primitive ideas used by the young child to think about space can be described as 'topological' concepts. Thus, it is claimed that children rely on such features as order (of points on a line) or enclosure (of points inside a closed curve), before Euclidean properties (such as length and straightness of lines or size of angles), in responding to and

identifying geometrical shapes. These conclusions are based
mainly on two sets of experiments, one set involving 'haptic
perception', in which children were asked to identify
different shapes by feel, and one set based on children's
drawings. Piaget and Inhelder also quote other studies to
support their claims, including, for example, one in which
children were asked to thread coloured beads onto a wire.

Piaget's experiments, and an early replication by Lovell
(1959), provide some evidence that children make use of the
above features, described as 'topological', before certain
Euclidean ideas; but this finding needs careful interpretation. It does not mean that all notions of a topological
nature are psychologically more primitive than any Euclidean
ideas. Moreover, although these concepts of order and enclosure may be made more precise in a mathematical system called
'topology', this does not mean that children can think in
terms of or in any way have access to the mathematical
system itself.

This confusion is exemplified in a nonetheless interesting study by Martin (1976b) who used items such as the one
below to measure whether children would choose topological
or Euclidean invariants when geometrical figures were
transformed.

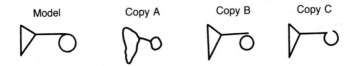

In the item shown, Copy A preserved the topological
features of the given material, whilst Copies B and C violated the notions of connectedness and closedness, respectively.
Martin argued that if topological actions preceded
Euclidean ones, children would have a marked preference for
Copy A. In the event, children tended to choose B or C. An
obvious explanation of this result is that B and C preserved
the straightness of the lines in the model, and, perhaps to
a lesser extent, the smoothness of the circle. Though these
aspects of a figure's shape can be regarded as Euclidean
notions (albeit extremely primitive ones), it hardly suggests
that connectedness and closedness are more advanced than the
Euclidean notion of congruence, say.

In one of Piaget's haptic perception experiments,
children are required to identify objects which they can
handle but not see by selecting copies from a set of drawings
or identical models. Children as young as six months can
distinguish between, say, a triangle and a circle, presented

150

visually; by hiding the figures in the present experiment, children are required to make mental representations - in other words it is their *concepts,* rather than percepts, of the figures that are being investigated. Lovell (ibid.) replicated this experiment and confirmed Piaget's finding that straightsided shapes such as a square, rhombus, trapezium etc., are the hardest to identify. Though these can be classified as Euclidean, Lovell argues that 'there is little evidence to suggest that it is topological properties, as such, which enable a child to identify certain shapes more easily than others'. In particular, Piaget and Inhelder seem not to have recognized that *curved* Euclidean shapes (such as a circle or ellipse)[?] are identified as easily as topological ones (such as an irregular shaped piece with a hole in it, or an open ring). Thus, Lovell suggests, it is the 'gaps, holes, curves, points, corners, ins and outs etc., in Euclidean space (that) make identification easier' (p.113).

An interesting side-line is provided by Lovell, and also by Fuson and Murray (1978), who compare children's ability to draw (straight-edged) shapes with their ability to construct them with matchsticks. In both studies, the construction task is found to be easier: Lovell's results suggest that construction precedes drawing by at least six months. The figure below shows all the shapes used by Lovell in the construction task, in order of difficulty (145 children, mean age 4:05 years).

These, together with other shapes shown below, were used by Lovell in his replication of Piaget's 'drawing task', in which children were asked to copy these shapes, presented individually on post-cards. The same children (N = 145, mean age 4:05 years) were involved as in the construction task.

Lovell's findings agreed with Piaget to the extent that, taking the figures as a whole, children seemed to find topological properties easier to convey than Euclidean ones (in relation to the marking scheme used). However, if only curved figures are considered there was no significant difference between the drawing of topological and Euclidean properties. More specifically, whilst Piaget maintains that children below the age of four years ignore Euclidean properties entirely, Lovell found that only 16 of the 43 children below this age drew 'rather similar shapes' for the circle, square and triangle. Also, these 43 children found the square and rectangle only a little more difficult than the 'topological shapes' 1, 2 and 3, whilst the circle was easier.

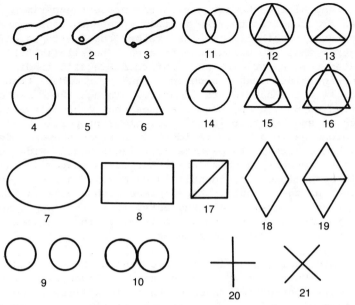

It seems clear that some topological notions are understood earlier than Euclidean ones, but the view that the child's first spatial concepts are exclusively topological is misleading and restrictive. It may be more appropriate to identify a progression from a reliance on simple qualitative cues to the recognition and application of increasingly complex spatial relationships. This progression involves a transition from global recognition of an entire shape to the analysis and appreciation of its component parts and properties.

There are a number of studies which suggest that children's understanding of shape and space becomes progressively more quantitative and analytic (e.g., Küchemann, 1980, Perham, 1978, Schulz, 1978, Thomas, 1978). For example, on analysing children's responses to a reflection task like the one illustrated in the adjacent diagram (in which children were asked to place on the right-hand triangle a penny, corresponding to the point A, Thomas (1978) found that the most common error made by six-year-olds was to locate the image of Point A on the correct side of the triangle but away from the nearest vertex. For eight-year-olds, one-third of the errors were also of this type, but another third consisted of placing the image-point directly on this vertex. In contrast, ten-year-olds made virtually no errors of any type. Thus, in this example, there seems to be a progression from focusing on one qualitative cue (side), to a less direct one, and one that involves quantity (nearest vertex), to a co-ordination of both cues.

In Küchemann's (1980a) report of the finds of the CSMS work
on reflection and rotation, a transition from global (qualitative) to analytic (quantitative) handling can be identified.
At the global level, the object is considered as a whole and
reflected as a single entity (not necessarily precisely, but
without overt errors). There is a transitional stage, at
which one point or a part of the object is accurately
reflected and the object then drawn in: sometimes at this
stage the resulting image has a less than plausible appearance. The final stage is the co-ordination of the analytic
elements of the task with the global appreciation to produce
a result which is accurate and looks correct. While this
transition is here described in terms of the strategy
adopted by children in thinking about the reflection and
drawing the image, it also features in their explanations and
answers to questions like 'How do you find the image?' and
'How do you know this is right?', hence appearing as a change
in the level of argument or justification adopted. Again,
this will be mentioned later.

Of the 14-year-olds tested by Küchemann (N ≙ 1000), well
over 90 per cent could cope *qualitatively* with items like
A1.1 and A1.7 (below), in the sense that their (drawn)
answers were not necessarily very precise, but equally the
answers contained no 'overt errors' (such as drawing the

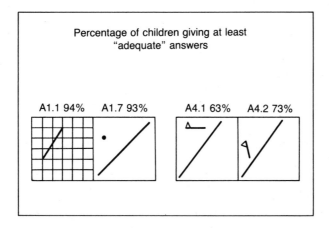

image parallel to the object or displacing the object
horizontally when the mirror line is slanting). On the other
hand, substantially fewer children gave qualitatively
adequate answers to A4.1 and A4.2.

To produce such answers for the first two items it is
only necessary to recognize that a reflection takes the object
'directly' across the mirror-line (A1.7) or that the image
'slopes the other way' (A1.1), which are both ideas that can
be derived very directly from the action of folding. For
A4.1 and A4.2, on the other hand, it is necessary to use and
co-ordinate both ideas, and, in particular, to relate the

slope of the object to the slope of the mirror-line - rather
than just to an external reference frame (e.g., the flag is
'horizontal' or slopes 'to the right') which leads to errors
of the type shown in the adjacent
diagram for item A4.1.

Responses where the slopes are not
coordinated

These pairs of items also differ
substantially in the degree of
control required to answer them
'accurately'. The first pair

(which had facilities 70 per cent and 61 per cent) can be
solved by a step-by-step approach (e.g., controlling the
direction of the displacement of the point in A1.7 and then
controlling the distance). In contrast A4.1 and A4.2 (which
both had facilities of 30 per cent) would seem to require a
far more analytical approach, whereby different elements of
the flag (the end points) are reflected separately, (rather
than building up the image by locating first one end-point
and then drawing the stem of the flag as can be done in
A1.1).

     Other studies of transformation tasks once more identify
the effects of context and complexity on the level of
difficulty.   Schulz (1978), for example, working with six to
ten year olds, confirmed that
transformations are easier when
the displacements are horizontal
or vertical rather than diagonal.
She also found that a familiar
object (the ship in the adjacent
diagram) was easier than an
unfamiliar one made from the same
elements, and that short displacements (20 cm) were easier
than long ones (300 cm).

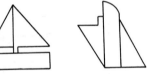

     Those groups of studies which identify developmental
changes in children's thinking about and operating on
geometrical tasks provide some understanding of why and how
younger children fail to carry out some such tasks satisfactorily. By the identification of those features which make
the tasks easier or more difficult, tasks can be arranged in
a suitable order; but, at the same time, if attention is
drawn to certain specific properties and aspects of geometrical configurations, it may be possible to accelerate the
transition to a more advanced level of understanding.

     In the context of geometry, however, certain curriculum
questions remain. Over the last twenty years, transformation
geometry has had widespread acceptance as a major topic in
school mathematics; but there is some evidence (see CSMS book,
1980) that this is in spite of a lack of conviction on the
part of teachers. It is worth pointing out that the
appropriateness of transformation geometry depends on the
objectives which it is intended to achieve. For most children,

154

it is unlikely to foster appreciation of a deductive system, any more than traditional Euclidean geometry did (indeed, this has not been attempted - see later section). Most children are unlikely to be able to obtain insight into a precise mathematical structure (particularly the group structure) through the combination of transformations; and, generally, O-level syllabuses demand this only in an incomplete and confused way.

The research evidence suggests, however, that the study of transformations can provide a considerable challenge to most secondary school pupils and can enhance children's understanding of, and ability to operate with and discuss geometrical relationships in a variety of ways.

It is worth mentioning the items of a geometrical nature in the APU surveys. The Primary Survey includes questions about angles, and names and properties of shapes, as well as more 'modern' topics such as symmetry and transformations. Examples of both are given in figures 3.2, 3.4 and 3.24. Amongst the 15-year-olds in the APU Secondary Survey, 75 per cent selected the largest angle from a set of five, compared with a facility of 65 per cent when pupils in the Primary Survey were asked to select the middle angle out of five.

Both the Primary and Secondary Surveys asked pupils to draw all the lines of symmetry on a capital 'E", a concave quadrilateral and a capital 'L'. The facilities were 55 per cent, 65 per cent and 50 per cent respectively in the Primary Survey, and 65 per cent, 80 per cent and 65 per cent in the Secondary Survey.

7. MEASURES

At present, school mathematics, especially at primary level, includes a substantial amount of material relating to measurement in its widest sense. This is reflected in the APU Primary Survey (1980), which includes a considerable number of items under the headings: money, weight, time, temperature, length, area, volume and capacity. The CSMS test paper dealing with measurement is restricted to items testing the understanding of length, area and volume.

The APU Primary Survey results in this area provide a collection of examples and comments which show up differences in difficulty, but do not relate these to one another in any systematic way. Items which test ideas of coinage and money notation, for example, generally gained facilities of 70 per cent or more, but the introduction of halfpence reduced the success rate to below 50 per cent. Some items about the use of a balance showed an object being weighed. When the object and the weights were on opposite sides, around 90 per cent gave correct answers, but over 30 per cent were confused when weights appeared on both sides, some pupils adding them

Items from lines, angles, shapes
Item cluster: angles
Mean sub-category score 48 per cent.

Response analysis		Item facility	
1% 3% 6% 76% 8%			P1 Put a ring round the angle which is a different size from all the others.
Response unclear	4%	76%	
Omitted	2%		
11% 13% 63% 4% 6%			P2
Response unclear	0%	63%	If these angles were arranged in order of size, which one would be in the middle? Put a ring round the middle sized angle.
Omitted	3%		
Incorrect	50%	44%	P3 Angle B is 60° Angle C is 40° Work out the size of Angle A.
Omitted	6%		
Incorrect	52%	36%	P4 What is the size of angle a?
Omitted	12%		
Incorrect	65%	15%	P5 The two lines with arrows on them are parallel. Mark two other angles on the diagram which are the same as the angle already marked.
Omitted	20%		

156

Items from symmetry, transformations and coordinates
Item cluster: line symmetry
Mean sub-category score 52 per cent.

Response analysis		Item facility	
Incorrect	16%	79%	B1 Draw the reflection of the shape in the mirror.
Omitted	5%		
Incorrect	15%	66%	B2 Draw in the line of symmetry on the shape.
Omitted	19%		
a) Incorrect	60%	19%	B3 Draw in all the lines of symmetry on these shapes.
Omitted	21%		a)
b) Incorrect	26%	50%	b)
Omitted	24%		
Incorrect	81%	14%	B4 Draw the reflection of the ⌐ shape in the mirror.
Omitted	5%		

Sub-category: modern geometry
Item cluster: transformations
Mean sub-category score–27 per cent.

	Item facility	Response analysis	

B1 — Item facility **73%**

Which two of the following shapes could be fitted exactly on top of each other if you cut them out?

B & D
....

| Incorrect | 24% |
| Omitted | 3% |

B2 — Item facility **56%**

We can describe translations on this page by using pairs of numbers.

The first number in any pair says how many steps east to take while the second says how many steps north:$\binom{2}{1}$ means 2 steps east 1 step north.

The grid is marked out in steps.

(i) Starting at 0 show on the grid the final position after the translation $\binom{-3}{0}$. Mark this point with an A.

| Incorrect | 31% |
| Omitted | 13% |

B3 — Item facility **38%**

(ii) Starting at 0 show the final position you would arrive at after the sequence of transformation₄ $\binom{4}{1}$ followed by $\binom{0}{-2}$ followed by $\binom{-1}{0}$. Mark this point with a B.

| Incorrect | 47% |
| Omitted | 15% |

Items from length, area, volume, capacity
Item cluster: area
Mean sub-category score–54 per cent.

Response analysis		Item facility		
Incorrect	27%	70%	D1	How many of the triangles fit into the rectangle?
Omitted	3%			
Incorrect	38%	57%	D2	Put a ring round each of the two shapes which have equal areas.
Omitted	5%			
Incorrect	45%	47%	D3	The area of this rectangle is 20 square centimetres. What is the area of the shaded part? ....square centimetres
Omitted	8%			
Incorrect 12 cm^2	23%	37%	D4	What is the area of this square? ....cm^2
Other incorrect	30%			
Omitted	10%			
Incorrect	54%	26%	D5	What is the area of the shape? ....cm^2
Omitted	20%			

D1:
4 cm, 3 cm, 1 cm, 1 cm

D4: 3 cm

D5: 5 cm, 1 cm, 1 cm, 2 cm, 2 cm, 4 cm, 4 cm, 1 cm

together and others ignoring those weights on the same side as the object. Questions testing the recording of time on the clock face had a facility varying from 60 per cent to 80 per cent: those concerned with duration of time-intervals generally came between 30 per cent and 60 per cent.

The above remarks all provide some clue about the features which make measure items more difficult. In the CSMS work, some attempt is made to identify general principles. Both the APU and CSMS results show that the introduction of a fraction or any other complexity makes any item very much more difficult. It is also true that children use naive methods (such as counting squares for an area) for longer than one would expect, and when these methods become cumbersome they are at a loss how to proceed. In the APU survey, high scores (57 per cent - 70 per cent) were obtained for items displaying plane shapes divided up into smaller congruent shapes. Only 37 per cent, however, were able to calculate the area of a shape as familiar as a square, given the length of one side. This length was given as 3 cm (small enough for the unit squares to be easily visualized), but 23 per cent gave the answer 12 cm, suggesting they were confused between area and perimeter. In many cases, formulae which have been taught are not readily avilable to the pupils; it is suggested in the CSMS book (1980) that they may have been taught just at the time when they were not needed, when the children were solving area and volume problems by counting.

The essential components of measurement considered in the analysis of the CSMS test items are:

(a) The length, area and volume of objects are not changed by displacements;
(b) Measurement can be quantified by the repetiton of a unit of measurement but the resultant number depends on the size of the unit used.
(c) Formulae for regular figures are short-cuts for the counting methods.

The way in which certain test items 'fool' children suggests that anything up to 30 per cent are not entirely convinced of (a). A remarkably large proportion are not convinced that the diagonal of a square is longer than its side, as the following item shows:

The 8-sided figure A is drawn below on centimetre square paper.

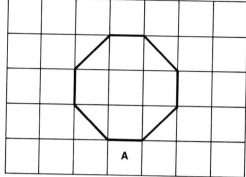

Draw a ring round the correct answer:

The distance all round the edge of A is:

8cm   more than 8cm   less than 8cm   you cannot tell

Table 2.1   Responses to Question 6

Answer	12 yrs	13 yrs	14 yrs
8cm	43.2	43.2	41.6 per cent
more than 8cm	38.5	36.7	46.9
less than 8cm	13.6	14.9	9.1

Again, a large number can easily forget the importance of the unit used to measure, particularly in the following:

John measures how long paths A and B are, using a walking stick. Then he measures how long paths C and D are, using a metal rod.

The answers are:

    Path A: 13 walking sticks

    Path B: 14½ walking sticks

    Path C: 15 rods

    Path D: 12½ rods

Draw a ring round the answer you think is true in each question:

a) Path B is longer than Path A: True/False/Cannot Tell

b) Path C is longer than Path B: True/False/Cannot Tell

c) Path D is longer than Path C: True/False/Cannot Tell

Table 2.2.   Responses to Question 3

	12yrs	13yrs	14yrs	
a) Correct	91.1	93.7	96.2	per cent
b) Path C is longer than Path B: True	50.9	33.8	27.3	
False	7.7	7.7	6.2	
Cannot tell	37.9	56.8	65.4	
c) Correct	95.9	92.8	94.1	

The interviews on this item showed again the child's dependence on a number answer, ignoring the unit of measure. The following conversation is taken from an interview on part b) of question 3 above where the child completely ignores the unit used to measure.

Paul (aged 12.9)	'Yes, it's got more'
Interviewer:	'What's got more?'
P:	'C's got 15 and B's got 14½'
I:	'C's got 15 what?'
P:	'Rods'
I:	'What about B?'
P:	'It's got 14½. Sticks'
I:	'So?'
P:	'C's longest'.

All the results quoted here provide evidence that measurement skills, supposedly acquired at primary level, cannot be taken for granted in secondary schools, and need constant reinforcement. In particular, it is clear that formulae and standard measurement procedures are valuable, but their appropriate use and retention is dependent upon both earlier practical experience and on a perceived need for and appreciation of their efficiency.

References

ADI, H. (1978): 'Intellectual development and reversibility of thought in equation solving', *J Res Math Ed,* 9, 204-13.

ASSESSMENT OF PERFORMANCE UNIT (APU) (1980): *Mathematical Development - Primary Survey Report No. 1.* London: HMSO.

ASSESSMENT OF PERFORMANCE UNIT (1980): *Mathematical Development - Secondary Survey Report No. 1.* London: HMSO.

BELL, A.W. (1979): *The Nature of Mathematical Learning: Some Comparisons with language*. Proceedings of the European Cognitive Research and Science Education Research Seminar, School of Education, University of Leeds.

BELL, A.W., SWAN, M.B. and TAYLOR, G.M. (1981): 'Choice of Operation in Verbal Problems with Decimal Numbers', *Ed Stud Math*, 12, 399-420.

BISHOP, A.J. (1980): 'Spatial abilities and mathematics education', *Ed Stud Math*, 11, 257-70.

BORKE, H. (1975): 'Piaget's mountains revisited: changes in the egocentric landscape', *Dev Psych,* 11, 240-3.

BRAINERD, C.J. (1979): *The Origins of the Number Concept*. New York: Praeger.

BROWN, J.S. and BURTON, R.B. (1978): 'Diagnostic models for procedural bugs in basic mathematical skills', *Cog Sci,* 2, 155-92.

BROWN, P.G. (1969): Tests of Development in Children's Understanding of the Laws of Natural Numbers. M.Ed. thesis, Manchester.

BROWNELL, W. and STRETCH, L. (1931): *The Effect of Unfamiliar Settings on Problem-Solving*. Duke University Research Stud Ed (1), Durham, N.C.: Duke University Press.

BRYANT, P.E. (1974): *Perception and Understanding in Young Children*. London: Methuen.

BURTON, R.B. and DEBUGGY (1980): 'Diagnosis of errors in basic mathematical skills'. In: SLEEMAN, D.H. and BROWN, J.S. (Eds): *Intelligent Tutoring Systems*. Academic Press.

CALDWELL, J.H. and GOLDIN, G.A. (1979): 'Variables affecting word problem difficulty in elementary school mathematics', *J Res Math Ed,* 10, 323-36.

CARPENTER, T.P. (1980): The Effect of Instruction on First Grade Children's Initial Solution Processes for Basic Addition and Subtraction Problems. Paper presented at the annual meeting of the American Educational Research Association, Boston, April, 1980.

CARPENTER, T.C. and MOSER, J.M. (1979): *An Investigation of the Learning of Addition and Subtraction*. Madison: Wisconsin Research and Development Center, University of Wisconsin.

COLLIS, K.F. (1975a): *A Study of Concrete and Formal Operations in School Mathematics*. Australian Council for Educational Research.

COLLIS, K.F. (1975b): *The Development of Formal Reasoning*. University of Newcastle, N.S.W.

COLLIS, K.F. (1978): *See* KEATS, COLLIS and HALFORD: *Cognitive Development*. New York: Wiley.

COPELAND, R. (1974): *How Children Learn Mathematics*. New York: Macmillan.

COX, M.V. (1975): 'The other observer in a perspectives task', *B J Ed Psych,* 45, 83-5.

DAVIS, R.B., JOCKUSCH, E. and McKNIGHT, C. (1979): 'Cognitive processes in learning algebra: student strategies', *J Child Math Beh,* 2, 1, 127-45.

DONALDSON, M. (1978): *Children's Minds*. London: Croom Helm.

EKENSTAM, A. and NILSSON, M. (1979): 'A new approach to the assessment of children's mathematical competence', *Ed Stud Math,* 10, 41-66.

FALK, R., FALK, R. and LEVIN, I. (1980): 'A potential for learning probability in young children', *Ed Stud Math,* 11, 184-204.

FEHR, L.A. (1979): 'Hypotheticality and the other observer in a perspective task', *B J Ed Psych,* 49, 93-6.

FIRTH, D.E. (1975): A study of rule dependence in elementary algebra. M.Phil. thesis, University of Nottingham.

FISCHBEIN, E. (1975): *The Intuitive Sources of Probabalistic Thinking in Children*. Dordrecht: Reidel.

FREUDENTHAL, H. (1973): *Mathematics as an Educational Task*. Dordrecht: Reidel.

FUSON, K. and MIERKIEWICZ, D. (1980): A Detailed Analysis of the Act of Counting. Paper presented at the annual meeting of the American Educational Research Association, Boston, April 1980.

FUSON, K. and MURRAY, C. (1978): 'The haptic-visual perception, construction and drawing of geometric shapes by children aged two to five: a Piagetian extension'. In: LESH, R. and MIERKIEWICZ, D.: *Recent Research Concerning the Development of Spatial and Geometric Concepts*. Columbus, Ohio: ERIC.

GALVIN, W.P. and BELL, A.W. (1977): *Aspects of Difficulties in the Solution of Problems Involving the Formation of Equations*. Shell Centre for Mathematical Education, University of Nottingham.

GELMAN, R. (1969): 'Conservation acquisition: a problem of learning to attend to relevant attributes', *J Exp Child Psych,* 7, 167-86.

GELMAN, R. (1972): 'The nature and development of early number concepts'. In: REESE, H.W. (Ed): *Advances in Child Development and Behaviour (7).* New York: Academic Press.

GELMAN, R. and GALLISTEL, C.R. (1978): *The Child's Understanding of Number.* Cambridge, Mass.: Harvard University Press.

GONCHAR, A.J. and HOOPER, F.H. (1975): *A Study in the Nature and Development of the Natural Number Concept.* Madison: University of Wisconsin Research and Development Center for Cognitive Learning. Technical Report No.340.

GREEN, D.R. (1980): Chance and Probability Concepts in Secondary School Pupils. Unpublished document, CAMET, Loughborough University.

GROEN, G. and RESNICK, L.B. (1977): 'Can pre-school children invent addition algorithms?' *J Ed Psych,* 69, 645-52.

HART, K. (1978): 'The understanding of ratio in the secondary school', *Maths in School,* 7, 1.

HART, K. (1981): *Children's Understanding of Mathematics, 11-16.* London: Murray.

HASEMANN, K. (1980): Difficulties with fractions. *Ed Stud Math,* 12, 71-88.

HITCH, G.J. (1978): 'The numerical abilities of industrial trainee apprentices', *J Occup Psych,* 51, 163-76.

INGLE, R.B. and SHAYER, M. (1971): 'Conceptual demands in Nuffield O-level chemistry', *Ed Chem,* 8.

INNER LONDON EDUCATION AUTHORITY (1979): *Primary School Mathematics. 1: Mathematical Content. 2: Checkpoints. ILEA.*

JANVIER, C. (1978): *The Interpretation of Complex Cartesian Graphs: Studies and Teaching Experiments.* Shell Centre for Mathematical Education, University of Nottingham.

JERMAN, M. and MIRMAN, S. (1974): 'Linguistic and computational variables in problem solving in elementary mathematics', *Ed Stud Math,* 5, 317-62.

JERMAN, M. and REES, R. (1972): 'Predicting the relative difficulty of verbal arithmetic problems', *Ed Stud Math,* 4, 306-23.

KARPLUS, R. and KARPLUS, E. (1974): *Proportional Reasoning and Control of Variables*. Cambridge, Mass.: Division for Study and Research in Education, M.I.T.

KÜCHEMANN, D.E. (1978): 'Children's understanding of numerical variables', *Maths in School*, Sept., 23-6.

KÜCHEMANN, D.E. (1980a): 'Children's difficulties with single reflections and rotations', *Maths in School*, 9 2, 12-3.

KÜCHEMANN, D.E. (1980b): 'Children's understanding of integers', *Maths in School*, 9 2, 31-2.

LAURENDEAU, M. and PINARD, A (1970): *The Development of the Concept of Space in the Child*. New York: International Universities Press.

LESH, R. (1978): *Recent Research Concerning the Development of Spatial and Geometrical Concepts*. Columbus, Ohio: ERIC.

LOVELL, K. (1959): 'A follow-up study of some aspects of the work of Piaget and Inhelder on the child's conception of space', *B J Ed Psych*, 29, 104-17.

LOVELL, K. (1972): 'Intellectual growth and understanding mathematics', *J Res Math Ed*, 3, 164-82.

LOVELL, K. (1978): 'Concept development'. In: WAIN, G. (Ed): *Mathematical Education*. Van Nostrand.

LUNZER, E.A., BELL, A.W. and SHIU, C.M. (1976): *Numbers and the World of Things*. School of Education, University of Nottingham.

MALPAS, A.J. and BROWN, M. (1974): 'Cognitive demand and difficulty of GCE O-level pre-test items', *B J Ed Psych*, 44, 155-62.

MARTIN, J.L. (Ed) (1976a): *Space and Geometry: Papers from a Research Workshop*. Columbus, Ohio: ERIC.

MARTIN, J.L. (1976b): 'An analysis of some of Piaget's topological tasks from a mathematical point of view', *J Res Math Ed*, 7-24.

MOSER, J.M. (1979): A Longitudinal Study of the Effect of Number Size and Presence of Manipulative Materials on Children's Processes in Solving Addition and Subtraction Verbal Problems. Paper presented at the annual meeting of the American Educational Research Association, Boston, April 1980.

MOSER, J.M. (1980): *Young Children's Representation of Addition and Subtraction Problems*. Madison: Wisconsin Research and Development Center, University of Wisconsin.

NESHER, P. (1976): 'Three determinants of difficulty in verbal arithmetic problems', *Ed Stud Math,* 7, 369-88.

NESHER, P. and TEUBAL, E. (1975): 'Verbal cues as an interfering factor in verbal problem solving', *Ed Stud Math,* 6, 41-51.

NOVILLIS, C.F. (1976): 'An analysis of the fraction concept into a hierarchy of selected subconcepts and the testing of the hierarchical dependencies', *J Res Math Ed,* 7, 131-44.

ORTON, A. (1979): Studies of Understanding of Functions and Elementary Calculus. Unpublished papers, School of Education, University of Leeds.

ORTON, A. (1983): 'Students' Understanding of Integration', *Ed Stud Math,* 14, 1-18.

PAYNE, J.N. (1976): 'Review of research on fractions', In: LESH, R.A. (Ed): *Number and Measurement: Papers from a Research Workshop.* Columbus, Ohio: ERIC.

PERHAM, F. (1978): 'An investigation into the effect of instruction on the acquisition of transformation geometry concepts in first grade children and subsequent transfer to general spatial ability', In LESH, R.A. (Ed): *Recent Research Concerning the Development of Spatial and Geometrical Concepts.* Columbus, Ohio: ERIC.

PIAGET, J. (1952): *The Child's Conception of Number.* London: Routledge and Kegan Paul.

PIAGET, J. and INHELDER, B. (1951): *La Génèse de l'Idée de Hasard chez l'Enfant.* Paris: Presse Univ. France.

PIAGET, J. and INHELDER, B. (1967): *The Child's Conception of Space.* London: Routledge and Kegan Paul.

PIAGET, J., INHELDER, B. and SZEMINSKA, A (1960): *The Child's Conception of Geometry.* London: Routledge and Kegan Paul.

REES, R. (1973): *Mathematics in Further Education: Difficulties Experienced by Craft and Technician Students.* Brunel F.E. Monographs (5). London: Hutchinson.

RESNICK, L.B. (1980): 'The role of invention in the development of mathematical confidence'. In: KLUWE, R. and SPADA, H. (Eds): *Developmental Models of Thinking.* Academic Press.

ROMMETVEIT, R. (1978): 'On the relationship between children's mastery of Piagetian cognitive operations and their semantic competence'. In: CAMPBELL, R.N. and SMITH, P.I.: *Recent Advances in the Psychology of Language, Part A.* New York: Plenum.

ROSSKOPF, M.F. *et al.* (1971): *Piagetian Cognitive-Development Research and Mathematics Education.* Reston, Va: NCTM.

ROSSKOPF, M.F. (1975): *Children's Mathematical Concepts: Six Piagetian Studies in Mathematics Education.* Columbia University, New York: Teachers College Press.

SAAD, L.G. and STORER, W.O. (1960): *Understanding in Mathematics.* Oliver and Boyd.

SCHULZ, K.A. (1978): 'Variables influencing the difficulty of rigid transformation during the transition between the concrete and formal operational stages of cognitive development'. In: LESH, R.A.: *Recent Research Concerning the Development of Spatial and Geometrical Concepts.* Columbus, Ohio: ERIC.

SHIU, C.M. (1978): *The Development of Some Mathematical Concepts in School Children.* Shell Centre for Mathematical Education, University of Nottingham.

STEFFE, L.P. (1975): *Research on Mathematical Thinking of Young Children: Six Empirical Studies.* Reston, Virginia: NCTM.

STEFFE, L.P. and PARR, R.B. (1968): *The Development of the Concepts of Ratio and Fraction in the Fourth, Fifth and Sixth Years of the Elementary School.* University of Wisconsin: Madison.

SWAN, M.B. (1980): Unpublished working papers. Shell Centre for Mathematical Education, University of Nottingham. (See also Bell, Swan and Taylor, 1981.)

THOMAS, D. (1978): 'Students' Understanding of selected transformation geometry concepts'. In: LESH, R.A. (Ed): *Recent Research Concerning the Development of Spatial and Geometrical Concepts.* Columbus, Ohio: ERIC.

VAN DEN BRINK, F.J. (1975): *Strokenaanpak I en II (An Approach to Ratio by Bars I, II)*, Utrecht, IOWO.

VAN DEN BRINK, F.J. and STREEFLAND, L. (1979): 'Young children (6-8) - Ratio and proportion', *Ed Stud Math*, 10, 403-19.

VERGNAUD, G. (1980): *Children's Understanding of Multiplicative Structures.* Cognitive Development Research in Science and Mathematics, University of Leeds.

WHEATLEY, G. (1976): 'A comparison of two methods of column addition', *J Res Math Ed*, 7, 145-54.

WOOD, R. and BROWN, M.L. (1976): 'Mastery of simple probability ideas among GCE ordinary level mathematics candidates',
Int J Math Ed Sci Tech, 7, 3, 297-306.

YOUNG, A.W. and McPHERSON, J. (1976): 'Ways of making number
judgments and children's understanding of quantity relations',
B J Ed Psych, 46, 328-32.

YOUNG, R.M. and O'SHEA, T. (1981): 'Errors in Subtraction',
Cog Sci, 2.

Chapter Seven
Aspects of Teaching Method

1.   INTRODUCTION

To get a perspective from which to view the research on
teaching methods it may be helpful first to set up a suitable
model of the teaching situation so as to see what choices
have to be made by the teacher (and the textbook writer).
The mathematics from which the teacher builds his curriculum
will already exist in his own mind as a set of interconnected
conceptual structures, each with applications in various
familiar areas of experience. For example, he will already
know how multiplication relates to addition and to division,
that it is commutative, and that it applies to, say, the
total number of children in 5 classes, each of 30 pupils.
The teacher's first task is therefore to select those
conceptual structures which he wishes the pupils to acquire,

and within these to identify which are the most important aspects (either because of their key position in the conceptual structure or on account of their frequency of occurence in application). He needs to consider also which *facts* will need to be memorized so as to be available on immediate recall, which *skills* are needed at a high level of fluency and speed, and which *contexts* need particular treatment as those to which the material is most likely to be applied. The next task is to decide on the route to be taken through the material - which concepts to introduce first, whether to return in spiral fashion to the main concepts at intervals, what particular problems to use. (This assumes that the route will be determined by the teacher; but some schemes allow the learner a measure of control over his route through the material - see the section on learner control, below.) At this point there is a need for a psychological 'map' of the conceptual field, indicating where the points of high and low difficulty are, and, more particularly, what the pupils' existing conceptions of the subject matter are. Here the research of the previous chapter is relevant.

Another decision relevant to this teaching design is what general orientation, what view of mathematics, is to govern the activity. Will the material be presented in a deductive framework, new principles being deduced from known ones; or will the approach be that of offering problems for joint investigation; or that of the direct teaching, and the learning by the pupils, of knowledge and skills? This kind of question is treated in the next chapter, on Process Aspects of Mathematics. (The teaching of *skills* is treated in Chapter Five.)

The subject of this present chapter is the choice of methods by which to teach the material that has been selected. Direct teaching, or 'exposition', and 'discovery' are two modes of teaching which many teachers would recognize, even if they would probably claim to use a combination of these for most of the time. The SMP and Scottish courses occupy fairly distant points on this scale. The research to be reported shows that exposition is generally more effective in the short term, but discovery is better for longer-term retention and for transfer to new situations. The analysis of a skill into component sub-skills and the teaching of these cumulatively, with a mastery criterion for progress from each sub-skill to the next, is also shown to be effective. But more effective methods still are those based on 'cognitive conflict'. These see learning as the breaking out of a limited framework of knowledge into a more inclusive one, and this being provoked by meeting situations which appear contradictory in the more primitive framework. Teaching by counter-example is one form of this.

The use of structured materials, the value of *immediate* feedback of correctness of answers, and the optimal placing

of periods of revision have also been studied and relevant
experiments are discussed below.

Many of the experiments described here achieved
spectacular success, and some of the methods developed in
them make our usual classroom practices look primitive. But
they have generally operated under more favourable conditions
than exist in the typical classroom. Nevertheless, there
would appear to be substantial benefits to be gained if we
could develop some of these methods into classroom form and
spread the knowledge of them more widely.

2.   EXPOSITION/DISCOVERY AND MEANINGFUL/ROTE LEARNING

The relative merits of discovery - compared with reception-
learning have formed a dominant theme in research on teaching.
However, there has been some confusion between this dimension
and that of meaningful - against rote-learning - not
surprisingly, since one of the main aims of discovery learning is to ensure that the learning is meaningful, that is,
well-integrated into the learner's existing knowledge.
Shulman (1970) and Ausubel (1968) point out that it is
possible to *discover* (e.g., by searching in a book) facts
which are intrinsically non-meaningful and, on the other
hand, that material can be assimilated meaningfully into
existing knowledge even if learnt from an expository presentation. Ausubel would add that this assimilation still needs
to take place even if the material is first obtained by
discovery. Ausubel (1968), in a long and useful chapter,
traces the development of the idea of learning by discovery;
he is in general critical of the research in its favour, but
concludes with the concession that it does have a place in
the learning of generalizations. (This is, of course, what
a very large part of *mathematical* learning consists of.)
The early work mentioned by him includes that of Katona (1940)
and Hendrix (1947). Katona, working with adolescent subjects,
used problems of the type shown below:

Change the 5 squares to
4 by moving three
matches.

He showed, among other results, that a rote-memorized
principle is less transferable to new problems than is mere
experience of problems exemplifying the principle. Hendrix
found that pupils were much more capable of applying a
principle where this had been learnt by experience of its
instances than where it had been verbally communicated.

In this country, Skemp (1971) reports two relevant
experiments. The first of these concerned the learning of
two 'Red Indian' type sign-languages; one of these had been
built up 'schematically', the meaning of a complex sign

depending on those of its constituent parts, the other being
generally similar but not fitting the learnt scheme, thus
having to be learnt by rote. Learning of the schematically
related language was very markedly superior to the other one,
particularly with respect to retention after four weeks.

% recalled	Immediate	After 1 day	After 4 weeks
Schematic	69	69	58
Rote	32	23	8

The other experiment was on the transfer of a rule for
traversibility of a network by a single path to cases where
the paths returned to their starting points. Those who
learned the underlying principle transferred more successfully than those who merely learnt a rule.

Discovery

As an example of the two ways in which mathematical generalizations may be discovered, consider the principle that the
angles of a triangle make 180°. This can be discovered by
measurement of a number of actual drawn or cut-out triangles.
It might also be discovered by experimenting with triangles
made with a rubber band on a pin-board (diagram below) and
observing that in all positions,

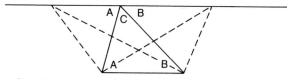

the two angles A are equal, as are the angles B, so that the
angle sum of the triangle is the same as that of the
straight line. The first method could be called empirical
discovery, the second deductive discovery. It is also
possible to discover concepts, though in a weaker sense,
that of becoming aware of the significance of a certain kind
of object and its properties. For example, out of a mixed
set of quadrilaterals the square might be noticed as having
a particular kind of regularity, and this seen to consist of
equality of sides and angles, these properties marking it off
from other figures similar but not identical, such as the
rectangle and the rhombus. This could take place before the
knowledge of any names for these shapes had been acquired.
In this sense the square might be discovered. It would of
course be appropriate for the teacher to supply the names at
this point. The names cannot be discovered intrinsically,
from the situation, but of course they could be found out by
the pupil without being told, for example by looking them up

in a dictionary. The aim of such an exercise might be to
train the pupil in general skills of learning. Similarly,
that 1 inch is 2.54 centimetres, although an arbitrary fact,
can be discovered by measurement of a line with two rulers.
The aim in this case might be to increase the pupil's
familiarity with the size of inches and centimetres.

Work by Land and Bishop (1967, 68, 69) at Hull
University interpreted discovery-teaching in terms of the
order Examples-Rule, as against Rule-Examples. They
explored the effect of this on learning, and also the effect
of using diagrams. This was with 12-year-old pupils.
Generally the order Examples-Rule was the more effective, but
the addition of the diagrams had a negative effect. The
reason for this may have been that the diagrams were in fact
offering a proof of the relationship (e.g., the area formula
for a triangle) after the relationship had been learnt; thus
it was answering a question which the pupils were not asking.
In subsequent experiments, diagrams were shown to have a
positive effect in the teaching of the generalization
$(a + 1)^2 = a^2 + 2a + 1$ (the bordered square diagram).

A thorough and more realistic experiment, conducted in
16 classrooms over a six-week period, is that of Worthen
(1968). The pupils were aged 11+, and the subject matter
concerned directed numbers and indices. The methods
differed chiefly in that under the expository method, the
principles to be learnt were stated by the teacher at the
outset of the lesson and examples were then solved by reference always to the stated rule; warning of likely mistakes
was given in advance. In the discovery method the pupils
were presented with a structured sequence of examples, but
with no hint from the teacher of an underlying principle; the
teacher acted as a helper for the pupils, but without the use
of any superior knowledge; if mistakes were made the teacher
left them until some pupil drew attention to them; pupils
were prevented from giving away the principle to others who
had not yet discovered it; the teacher did however state the
principle when all, or nearly all, of the class had
discovered it.

The eight teachers in this experiment underwent thorough
training; each taught two classes, one by each of the two
methods. Their conformity to the appropriate method was
monitored by observers and pupil questionnaires. The pupils
were tested immediately at the end of the six week period,
and five and eleven weeks later, on a number of measures.
The significance levels of the D/E differences are shown in
the table overleaf. (The heuristic transfer tests comprised
mathematical problems which could be solved by the discovery,
in each case, of an underlying principle.) As can be seen,
the discovery treatment proved inferior on the immediate post-
test, but superior on the retention and transfer tests and
the tests of acquisition of problem-solving strategies

(heuristics). There were no significant differences in attitudes.

MEASURE	P	DIRECTION
Concept Knowledge Test	< .01	D < E
Concept Retention Test 1	< .05	D > E
Concept Retention Test 2	< .025	D > E
Concept Transfer Test	< .08	D > E
Neg. Concept Transfer Test	n.s.	
Sem. Diff. Attitude Scale	n.s.	
Statement Attitude Scale	n.s.	
Written Heuristic Transfer	< .05	D > E
Oral Heuristic Transfer	< 0.025	D > E

Roughead and Scandura (1978) asked what it was which accounted for the superiority of discovery-learning. They concluded that it was the learning of how to approach problems of the types concerned, and argued that these methods and approaches might be taught equally or more effectively by exposition. Their experiment showed this to be the case, but they add a caution that it is not always easy to identify the approaches being learned sufficiently clearly to teach them. We might add that one would also lose the possibility of development of attitudes of willingness to tackle problems for which strategies had not been taught.

The inhibiting effects of failing to discover were shown in another experiment of Scandura et al. (1969). The converse result, that *successful* discovery is strongly and consistently related to learning, was shown by Anthony (1973), who re-analysed the results of a number of earlier experiments and showed that their apparent inconsistencies arose from failure to separate successful from unsuccessful discovery.

To summarize, the claimed advantages of discovery-learning (shared to varying extents by these different types) are:

(1) it ensures meaningful learning, since the pre-requisite knowledge must be activated before the discovery activity can progress;

(2)   it presents situations in the same ways as those in
      which the learning will need to be used subsequently;

(3)   it promotes the learning not only of the principle
      itself but of general strategies for the investigation
      of problems,

(4)   *if the discovery is successful,* it is highly motivating.

These points all emerge in the research. The general
conclusion is that discovery is often less effective than
exposition for immediate learning, but is better for
retention and for transfer to new situations. This refers to
guided discovery. Pure discovery, in the triangle example
quoted above, might ask pupils simply to investigate the
angles of different triangles; this would be unlikely to be
very successful. In fact, the closer the guidance, the
greater the chance of successful discovery but the smaller
the opportunity to develop personal strategies of inquiry.
Later research has tended to move away from considering
discovery as such, towards a more detailed analysis of
learning tasks. Some recent work on the effect of meaningfulness appears in the section, later in this chapter, on
computer-based learning.

Individual differences and discovery

Some recent English research has shown that personality
dimensions may affect the suitability of discovery or
exposition for particular pupils. Trown (1970) in one
experiment in which twelve-year-olds were introduced to the
Midland Mathematical Experiment's approach to vectors, using
number pairs and coordinates, by (1) statement of rules
followed by their use in examples, or (2) examples used to
instigate search for rules, found that though there were no
significant differences overall, the extraverts 'were
severely handicapped by the worksheets which thrust rules
upon them before they had a chance to find out for themselves'. In a second experiment (Trown and Leith, 1975) in
ten-year-old classrooms, using normal primary school
discovery or expository methods to teach directed numbers
(by number pairs, following the Nuffield Guide, Computation
and Structure 4), it appeared that, although again over-all
differences were slight, highly anxious children did badly
under the discovery method, though not under the expository.
This anxiety effect appeared also in Bennett's (1976) comparison of formal and informal primary classrooms.

3.   SKILL INTEGRATION, USING LEARNING HIERARCHIES

This is Gagne's (1970) theory of instruction. The assumption
is that a complex task can be analysed into a hierarchy of
sub-tasks, and that failure to perform the complex tasks can
be traced to lack of competence in one or more of the sub-

tasks. These in turn can be similarly analysed until one reaches sub-skills which are within the learner's previous competence. One then takes the learner upwards through the resulting hierarchy, so that at each step he has only to put together two or more of his existing skills to achieve the new skill; this is practised until he achieves mastery of it. An example of such a hierarchy is shown in the diagram below. This was the subject of a teaching experiment by Trembath and White (1978). They obtained the hierarchy by analysing students' errors on a previous test. The teaching,

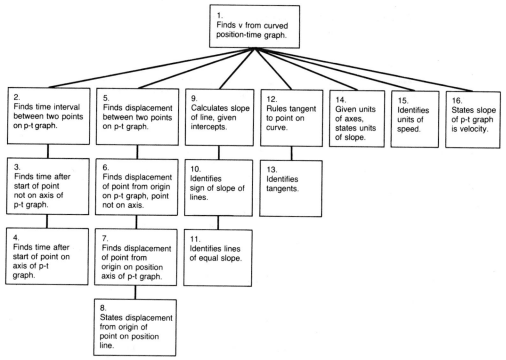

using individual booklets, lasted about one hour; the 13-year-old students involved were divided into four groups which received somewhat different treatments. Two groups worked forwards through the hierarchy; pupils in the first group passed from one task to the next when they were able to answer three successive questions correctly; pupils in the second group moved on after *one* correct answer. The other two groups began by looking at the final task, and coming backwards down each line of the hierarchy until they found a task they could perform; then they worked forward, again with either a three-question (R3) or a one-question (R1) mastery criterion for passing to the next task. The results are shown in the tables overleaf.

Numbers of Learners Attaining Criterion for Skill 1
in Learning Program

	Group			
	F3	F1	R3	R1
Number attaining criterion[a]	37	34	26	32
Number in group	39	37	34	38

[a] Criterion for F3 and R3 was three questions correct in
succession; for F1 and R1, one question correct.

Mean Times (Minutes)
Taken by Groups

	Group		
F3	F1	R3	R1
58	47	66	59

Numbers of Learners Correctly Answering Zero,
One, Two, and Three Questions in Post test

	Number of questions correct				
Group	0	1	2	3	Total n
F3	4	8	9	18	39
F1	13	8	7	9	37
R3	8	9	5	12	34
R1	20	6	5	7	38
Grade 11	42*	18	20	49	129

There was no direct comparison with another method,
although it was reported that the one hour's instruction
produced results superior to those of pupils three years
older who learnt the topic, as part of their normal curriculum, over a considerably greater time. However, there
was no test of retention, and it might well be that this
result achieved on a single skill out of context is unlikely
to be maintained over time. How do these hierarchies compare with normal classroom instruction? Breaking a complex
task into simpler tasks is normal practice; the differences
are in (1) the precise identification of the target task,
(2) the degree of detailed breakdown into sub-tasks, (3) the
mastery criterion, applied individually.

These aspects of the method are worth pursuing further.
But there are other aspects implicit and needing consideration. The analysis into sub-tasks involves judgments
about how fine the subdivision needs to be and it ignores
the question of whether it is easier to learn a new skill

when it is presented not as an isolated task but in the
context of the target task.

4. ADVANCE ORGANIZERS

A principle of Ausubel's (1968) theory is that meaningful
learning is facilitated by the use of an *advance organizer*.
This is an introductory overview which indicates the position
of the new material in the framework of knowledge already
possessed, and provides higher-level organizing concepts.
In one of Ausubel's experiments, the learning of the properties of steel was preceded by an introduction comparing
metals with alloys, and discussing their advantages and disadvantages. Another experiment taught the characteristics
of Buddhism with and without an introduction comparing
Buddhism with Christianity. Both of these experiments
showed the effectiveness of the advance organizer (Ausubel,
1968). In two experiments based on college-level mathematics, Lesh (1976) showed significant effects resulting
from the use of an introductory organizer consisting of an
intuitive overview of the subsequent mathematical material,
with the discussion of concrete and familiar models of the
structure. The topics were (a) finite geometry and (b)
groups. In the first experiment two different teaching programmes were employed, one based on a learning hierarchy
(see above), presenting the concepts in a standard deductive
sequence, while the other used a spiral approach in which
the most important ideas were presented several times at
progressively more sophisticated levels. The organizer in
this case explained informally the similarities and differences between finite and Euclidean geometry, and illustrated
them by pegboard pegs and rubber bands, elements (blocks) and
sets, members and committees, and others. The organizer
was given before the formal teaching to one group, afterwards
to another. The results showed that the use of the advance
organizer before the unit was effective. Lesh's second
experiment, dealing with finite groups, employed only one
teaching method (hierarchy) but compared two introductions.
Both gave an overview and discussed concrete examples, but
one emphasized examples of groups, the other non-examples
of different aspects of the structure, for example, the non-
commutative triangle group, and some non-associative and
identity-lacking systems based on Cuisenaire rods. The
results showed the superiority of the advance over the post-
organizers, and a smaller advantage for the counter-examples
introduction over the other.

A number of experiments on advance organizers have been
carried out, in different subject areas. The results have
been mixed, partly due to some confusion about the precise
nature of an organizer; a discussion of the controversy has
taken place in the pages of the *Review of Educational
Research* (Barnes and Clawson, 1975; Ausubel,1978; Lawton and

179

Wanska, 1977). One point at issue is the assertion that the organizer should be material at a higher level of abstraction and generality than the material to be learned. Lesh's organizers would not entirely fit this definition, though they do accord with the looser descriptions given by Ausubel elsewhere (1968), in comparing the new concept in general terms with one already known.

Lawton has conducted a series of experiments with children of primary school age which focused strongly on 'higher level' organizers. One of these used a treatment which emphasized the *general properties* of classification, ordering and numeration before a teaching sequence involving the use of these operations with concrete objects. This was successful.

Other experiments on the learning of classification skills compared organizers consisting of (a) higher order concepts by which the subsequent material might be organized (e.g., concepts of food, shelter, tools for organizing information about the lives of men in ancient and modern times), and (b) instruction in *how to classify* (e.g., pick a number of pictures and consider how they might go to-gether). A third experimental treatment (c) combined (a) and (b). The order of effectiveness was $c > b > a$ (Lawton and Wanska, 1979).

How far can this work be applied to the mathematics classroom? First, our knowledge of long-term memory (see Chapter Two) indicates that richly connected material is the best retained, and though hierarchical classification is by no means the only form of organization by which material is stored in the memory, it is a particularly useful one to develop. This suggests that the use of organizers should be explored. But, secondly, the use of a *higher level* organizer may not be possible without first teaching this more abstract set of concepts. This was an important part of Lawton's teaching programmes; while the non-availability of the general notion of an axiom system precluded Lesh from using this as an organizer; he had therefore to use a *comparative* organizer, relating the new idea (finite geometry) to one of the same level (Euclidean geometry), with the use of lower-level exemplars (committees, etc.). Thirdly, mathematics lessons often begin by discussing the type of problem which the new idea or method is intended to solve, sometimes as a deliberately unexplained situation to provoke curiosity, sometimes introducing this by previous work in the same field. This latter is clearly also a valid way of establishing connections in the structure of knowledge, but it is possible that these connections, which may be contextual more than structural, are less beneficial ones to promote than the classificatory ones (see Paige and Simon, 1966; Silver, 1978). The general conclusion would be that considerably more could probably be done to help pupils to orient themselves to the kind of concept or method being taught, and also to build up a better connected system of knowledge, by discussing material qualitatively in terms of higher order concepts. It is possible, too, that the basic processes of classifying and establishing isomorphisms and symbolizations could be more explicitly used to help pupils understand the nature of mathematical activity.

5. THE PIAGETIAN TEACHING METHOD - COGNITIVE CONFLICT

It is often asserted that Piaget's aim has been to describe and explain the development of the intellect, and not to investigate how its development might be accelerated by specific teaching. But an important set of teaching experiments on the conservation of number, quantity and length and on class inclusion, were conducted at Geneva by Inhelder, Sinclair and Bovet, and the results published in 1974. These were actually designed, not simply to achieve maximal learning, but rather to expose more clearly the child's natural learning processes by attempting to reproduce a similar process by teaching interventions. However, they do clarify what kinds of teaching method the Genevans regard as consonant with their understanding of development. The authors state

'We started with the idea that under certain conditions an acceleration of cognitive development would be possible, but that this could only occur if the training procedures in some way resembled the kind of situations in which progress takes place outside an experimental set-up.' (op. cit., p.24).

Thus the Piagetian teaching method, like the skill-integration method of Gagné, requires prior experimentation, in this case to discover the *normal stages of development* through which children pass in acquiring the concepts under consideration. The prior experiments of the skill-integrationists aim to analyse the task to determine what prerequisites are needed for the performance of the final task, so that these may be directly taught.

The main difference is in the depth of the task analysis. The Gagné method starts by observing what steps are involved in the successful performance of the final task, and what breakdowns occur. It then teaches this sequence of steps, with warnings and practice designed to avoid the mistakes. The Piagetian method tries to recognize the conceptual system in which the unsuccessful performer is operating, and how this differs from that of the successful child. Experiences are then designed in which the use of the inadequate conceptual schemes leads to conflict between contradictory results; the child is thus stimulated to seek resolution, which involves restructuring his conceptual system, so that he can then deal with the task in question. The Piagetian method involves mental activity which may or may not be based on physical manipulation of materials; it is essentially *reflective* activity, not the learning of a skill. The Piagetian method also demands strong criteria for accepting that learning has taken place; the child needs to be able to explain his conclusion, to resist counter-arguments, to transfer his knowledge to new tasks differing both in small and in big ways from the learning task, including changes of context, and to retain these performances over a period of some weeks. Only then would the Piagetians recognize that genuine learning had taken place (Inhelder, Sinclair and Bovet, 1974).

A critique of the Piagetian approach *as a teaching method* was published by Brainerd (1978). He argues that the Piagetian method of the presentation of situations intended to provoke discovery without explicit feedback of correctness, is not as effective as more direct teaching methods, even when the strong criteria for learning - the demonstration of explanation, transfer and retention - are accepted. In one relevant experiment, Beilin (1965) taught number - and length - conservations by four methods. One of these consisted of making changes of the material without verbal comment - for example, the rearrangement of a set of matchsticks form making a connected straight line to a zig-zag (below), a second method did the same, but included also the direct teaching of the relevant rule, that is 'If I don't add or take away any sticks, but only move them, it stays the same length, even though it looks different'.

Under the former treatment 39 per cent of the trained group of six-year-olds passed a post-test on length conservation, and with the direct rule instruction 70 per cent succeeded; but there was little or no transfer to an area conservation task requiring the recognition of equality of area of two different shapes formed from the same number of unit squares (below).

In another experiment, Gelman (1969) trained five-year-old non-conservers on number and length using 32 problems. Two of these are shown schematically below; they were

actually presented using objects. The other problems were
variations of this with repect to colour, size, shape,
starting arrangement and combinations of quantity.

PROBLEM TYPE

Trial	Number	Length
1	•• •• •• / •• ••	═══ ─
2	•• •• •• / •• ••• •	── ═ ──
3	•• •• •• / ••• •• •	── ═ ──
4	••• •• / • •• •• •	── ──
5	•• • • / • • ••	─ ═ ─
6	••••• ••••• ••• / •••••	── ── ──

For the length training, in each trial children were
asked to point to two configurations which had the same
length and to two which had different lengths. They were
told immediately 'Yes, that's right' or 'No, that's not
right', but nothing more; the correct choice was not indicated. 16 six-trial problems were presented to the children
on each of two successive days, length and number problems
alternately. On the third day the post-test was given,
consisting of conservation tests for each of the concepts
number, length (trained), liquid quantity and substance
(untrained). The post-test was repeated two to three weeks
later. Two control groups were used, one of which had the
same problems but without feedback of correctness, the other
using problems in which the differences between the configurations were obvious and qualitative, not quantitative,
for example two toy lions and one toy cup. The results
showed that the trained children produced 90-96 per cent
correct responses in the post-tests of number and length and
about 55 per cent correct in the transfer to liquid quantity
and substance. The non-feedback group scored 20-30 per cent
on number and length, and virtually nothing on the transfer
tasks. The second control group scored almost zero throughout. In the delayed post-tests, the 90-96 per cent responses
were maintained, while the scores on the transfer tests
increased from 55 to about 70 per cent. It was also observed
that the quality of explanations improved between the two

183

post-tests.

Comparing these results with those of Beilin, who, with
verbal rule training, obtained less good results and little
transfer to other conservations, Gelman suggests that Beilin's
use of significantly fewer trials, and his presentation of
the number and length training separately instead of alter-
nating, were the factors responsible for the differences.
One may add that the presumably advantageous rule instruction
of Beilin as compared with the simple Yes-No feedback of
Gelman is shown to have less effect. An attempt to replicate
Gelman's experiment with a modified and more stringent post-
test confirmed the general results but obtained considerably
less learning (Eull and Silverman, 1970). A similarly
conceived experiment by Sheppard (1974), training for conservation of liquid quantity and substance using an array of
containers (or plasticine cylinders) varying in height along
one dimension and in width along the other, obtained about
40 per cent success on the trained conservations, and 20-30
per cent success on transfer to number, length and other
conservations. His training method consisted of a demonstration in which the children were asked to predict the outcome
of the experimenter's action (e.g., pouring liquid from one
container into another); the action was performed and the
outcome stated.

In these experiments the children were required only to
look at the situation and to respond. The Genevan experiments, by contrast, required the *manipulation* of the materials
by the children; also, no verbal feedback of correctness was
given, although following incorrect responses a fresh problem
was given which would induce conflict and so cast doubt on
the previous response. In these experiments, some of the
subjects were partial conservers at the outset. The success
rates ranged from 20 to 40 per cent; rather less than those
achieved by more directed learning. In recognizing the
effectiveness of the feedback and rule-training methods, one
should note that this was by no means the verbal exposition
of a principle without regard to whether its meaning was
understood, since the representation of the concept in
concrete form, with familiar materials, was the common
assumption in all these experiments. One may conclude that
provided this condition is met, learning may be provoked in a
number of effective ways; sufficient repetition and immediate
feedback of results are helpful. Genuine far transfer,
retention and understood explanations are the safeguard
against superficial learning; any method which shows success
on these criteria must be accepted as valid.

Two rather different types of instruction have also been
experimented with. In *observational learning* the subject
observes another person (live or filmed) performing the
criterion tasks; in *conformity training* he is grouped with
peers who already have the knowledge in question. Brainerd

(1978) quotes a number of such experiments: here we give just
one example of conformity training carried out by Murray
(1972):

> Children who fail pre-tests for a concept are grouped with
> one or more children who passed the same tests. During
> training, several administrations of the pre-tests are
> given, and the children are told that they may not give
> individual responses on these tests. They are instructed
> to discuss each test question and to formulate a consensual
> answer. After training, the children who originally failed
> the pre-tests are post-tested to determine whether or not
> their performance has improved. In Murray's original
> research, six conservation concepts (number, mass, space,
> liquid quantity, discontinuous quantity, weight) were
> subjected to conformity training in two experiments.
> Roughly 79 per cent of the pre-test nonconservers learned
> all five concepts. Roughly 81 per cent of the pre-test
> nonconservers showed transfer to two untrained conservation
> concepts (length and area). Both the judgment and
> explanation responses of Murray's subjects improved and all
> improvements were stable across a 1-week interval.
> (Brainerd, p.88.)

Such a report reminds one of pupils' considerable
predilection for wanting to know their friends' answers (and
methods?), and the dangers here of superficial learning and
presentation of non-understood answers. However, these
dangers occur in school situations where having the correct
answers acquires an extrinsic reward. Where learning success
is always judged by the strong criteria discussed above, and
pupils understand and accept these criteria, it may be
possible to exploit pupil interaction more positively.

6. NEO-PIAGETIAN TEACHING METHODS

Reference has been made in Chapter Three to theories which
ascribe Piagetian stage developments mainly to an age-related
increase in information processing capacity. Case (1975,
1978) has devised a method of instructional design based on
this theory and on the notion of field dependence (Chapter
Two). His initial assumptions are:

> (a) that young children tend to develop reasonable but over-
> simplified strategies for coping with the tasks they are set
> at school, and (b) that one reason they do so is that they
> are incapable of coping with the informational demands of
> the learning situation. (Case, 1978, p.442.)

These lead to the following procedure:

> 1. Precede the design of instruction with a step by step
> description not only of the strategy to be taught but also
> of the strategy or strategies that children apply to the

instructional task spontaneously.

2. Design the instruction in such a way that the limitations of children's spontaneous strategies will be apparent
and that the necessity of applying the strategy to be taught
will therefore become clear.

3. Both in selecting the strategy to be taught and in
designing the sequence of instructional activities to teach
it, reduce the working memory requirements of the learning
situation to a bare minimum.   (Case, p.442.)

The first two of these three principles are similar to
those of the Piagetian methods described above.   Important
details added by Case include (1) that the pupil must have a
means of checking for himself whether or not his answers are
correct, (2) he must come to see, if necessary through teacher
discussion, *why* his spontaneous strategy breaks down, and why
the correct strategy, which he is taught, is successful.   The
minimization of load on working memory implies (1) keeping
down the number of items of information to be attended to at
any one time, (2) practising each new sub-skill up to mastery
in its simplest form before adding complications, one by
one, and (3) deliberately emphasizing (at first, then less so)
the important but non-obvious features of the situation.
(A test for deciding which of two procedures carries the
higher memory load is to give them to untrained adults and
ask which they find takes the greater mental effort.)

Case describes the application of his method to the task
of separating variables , as, for example, in Piaget's Bending
Rods problem, where rods which differ in width, shape of
cross-section, material and length are given, and the factors
affecting the amount of bending produced within by a weight
have to be determined.   The expert strategy for this problem
involves selecting two rods to compare which differ in one
respect only, so that after choosing one rod, all four of its
characteristics have to be noted, along with its position,
while the remaining rods are scanned.   This implies carrying
five items in the working memory.   Young subjects normally
fail by picking two rods which differ in one respect, say
fat/thin, and comparing their degress of bending, without
noticing whether they differ in some other way as well.   This
is explainable as an inability to carry more than two items
in working memory - position and 'fat'.   A strategy which
might be taught could consist of continuing the spontaneous
procedure, when a thin rod is found, by checking another of
its attributes, returning to the first rod to see if it
matches, if so, considering a further attribute, and if not,
rejecting the second rod and seeking another which is thin,
then repeating the checks.   This changes the 'expert'
strategy by considering the different attributes sequentially
instead of simultaneously, and so drastically reducing the
memory load, in exchange for an increase in the amount of

scanning needed. The design of an instructional method for this procedure requires the provision of a recognizable means of checking the failure of the spontaneous strategy. In the Bending Rods problem the breakdown only becomes apparent when and if inconsistent results arise. Case dealt with this difficulty by using a teaching situation in which the variables to be separated could actually be physically separated. This was a rod and block task in which four rods, of two different materials (aluminium and brass), were contained in (pushed through) light and heavy wooden blocks (coloured light and dark). The problem was to find, by weighing on a balance, which type of rod ('silver' or 'gold') was the heavier. The learners (eight-year-olds) were given the problem, allowed to make a mistake (the light rod in a heavy block and the heavy rod in a light block were placed nearest to them), and then the rods were removed and weighed separately, and the heavy block shown to be responsible for the misleading result. The learners made further tests, with new rods, each time checking their own correctness by removing the rods. In a second training session a similar two-variables task in which the variables could be separated physically was used (hard and soft balls dropped from different heights), and in the third and fourth sessions three- and four-variable problems were introduced. The training took about 80 minutes in all. About 80 per cent of the subjects succeeded in the Bending Rods task after this training, compared with 20 per cent of a control group (who learned from the pre and post-tests themselves). This figure of 80 per cent of eight-year-olds is higher than the level obtained by untrained 15- and 16-year-olds. Moreover, a two month delayed post-test showed a further improvement, and further similar experiments have shown transfer to other tasks.

Gold's experiment on the teaching of proportion

These results represent a spectacular degree of acceleration of normal development of this aspect of logical reasoning, and suggest that the features added by Case to the Piagetian teaching methods previously discussed may have considerable significance. Experiments involving teaching standard curriculum topics are also quoted by Case, and one of them (Gold, 1978), in which a 'Case' method for the teaching of ratio and proportion was compared with a 'Gagné' method, based on skill integration, is described below.

Gold's experiment was aimed at solution of problems of the types:

'Joe can walk 6 miles in 2 hours. How far can he walk in 4 hours?'

'I can cook 2 pies with 5 big apples. How many pies can I cook with 12½ big apples?'

The pupils were (a) above-average 10- and 11-year-olds.
(b) below-average 12- and 13-year-olds.

The standard text-book method and the 'Gagné' method both attempted to teach, one by one, the steps of a correct solution of the form:

Joe's first walk rate $\dfrac{6 \text{ miles}}{2 \text{ hours}}$ (Identify rates)

Second walk $\dfrac{? \text{ miles}}{4 \text{ hours}}$ (Complete and incomplete rates)
(Write as fractions)
(Get them right way round)

$$\frac{?}{4} = \frac{6}{2} \qquad \text{(Equate the two fractions)}$$

$$2? = 6 \times 4 \qquad \text{(Cross multiply)}$$

$$= 24$$

$$? = 24 - 2 \qquad \text{(Divide)}$$

$$= \underline{12}.$$

Answer   <u>12 miles</u>   (Put in unit)

The Gagné method differed from the textbook in breaking down this process into 16 sub-tasks of learning a new step in the process or integrating it with previous steps; pupils were taught individually, and test questions were asked after each sub-task, with immediate feedback of results and reteaching before proceeding to the next task. This took an average time of about one hour. Two contexts were used throughout, a mixture problem ('This first jar contains 2lb of walnuts and 3lb of almonds; the second contains 4lb of walnuts how many pounds of almonds must it have to give the same mixture?') and a juice problem in which numbers of glasses of lemon juice and water were to be mixed to give drinks of the same strength. The substances and the containers were represented by cardboard rectangles labelled 'walnuts', 'jar' and so on.

The 'Case' teaching method was aimed at a solution method which did not use a specific algorithm, but used the key concept 'for every'; pupils were taught always to ask 'How many juices are there for every one of water?' or vice versa. So Joe's walk problem could be solved:

6 miles in 2 hours.
3 miles in 1 hour.
In 4 hours, 12 miles.

Or perhaps directly:

'4 hours is twice 2 hours; so twice 6 miles, i.e., 12 miles.'

The teaching method was based on a sequence of tasks each of the same form as the final task, but working up through nine levels of difficulty with respect to the numbers concerned; level 3 included 1 juice to 5 waters, 3 juices to ? waters; level 6 included 5 to 5, 3 to x; level 9 included 2 to 7, 5 to x. Again there was immediate feedback; the teacher asked the pupil to check any wrong answers by making up the mixture with the cardboard pieces, for example for the level 3 question, making up the three sets of 1 juice, 5 waters. Wrong answers were actually proposed to the pupil for consideration and demonstration, by him, of how they were wrong. Each set of problems was retaught twice, if necessary, before moving to the next level.

Two contexts were again used, the mixed nuts problem being replaced by one concerning boxes and pieces of gum, but the whole teaching sequence was taken through in one context (boxes of gum), before moving to the other.

The levels used were derived from previous research, as indeed were the Gagné sub-tasks, the difference being in the depth of analysis of the pupil's task and of his difficulties. The Gagne theory assumes that the pupil's difficulty is in remembering the steps of the correct solution method, and in getting and transforming the correct expression $2/6 = 4/x$, whereas the Case theory assumes that the number processing may be done in a variety of ways, depending on the actual numbers, but will be guided by the pupil's knowledge of what he is trying to achieve. It also assumes that the understanding of the problem structure will be best built up around the emphasized key concept, rather than by the teaching of a sequence of steps.

The results of the experiment showed the Case method to be superior to the Gagné, particularly for the below-average 12 to 13-year-old group. See the table below.

Group	10 - 11 above average	12 - 13 below average
Standard	33	11
Gagné	78	33
Case	100	100

Percentages achieving criterion 13/20 on delayed post test.

The success of the Gagné method with the 10 to 11-year-old above-average group was perhaps surprising. It may indicate that some of this group achieved spontaneously an understanding similar to that of the Case group; this is supported by the fact that some of them were observed to be using Case-type strategies in the delayed post-test. In our view, it shows how able pupils, given a procedure to follow, but relatively little explanation, can master the procedure sufficiently well to be able to reflect on it and to supply the rationale for themselves.

7. EMBODIMENTS

The teaching of abstract ideas through a particular model or embodiment of these ideas is a pervasive and familiar part of mathematics education. The use of a number-line for directed numbers, of shaded areas to show fractions, and the 'base blocks' for place value are examples.

Dienes (1960) stated two principles relating to this use of embodiments, based mainly on the knowledge and theory then available of children's thinking processes, but also to some extent on a series of experiments conducted in Leicester (Dienes, 1959). The *constructivity* principle stated that construction always needs to precede analysis: construction refers to the process of 'putting things together to build another structure with some previously specified requirements'. Dienes (1963) shows photographs of children working at the constructive stage of finding factors, arranging 35 blocks in rectangles of different dimensions. The *perceptual variablity* principle stated that to abstract a mathematical structure effectively, one must meet it in a number of perceptually different situations to identify its purely structural properties.

However, in the light of later work, Dienes modifies these statements. His 'An experimental study of mathematics learning' (1963) provides evidence that analysis of mathematical ideas can arise 'either out of construction or out of rule-bound play, but mostly out of the second'. These ideas are developed further when Dienes (1973) shows examples of teaching approaches through the systematic exploration of either a concrete embodiment or a game: these include activities with logic blocks, base blocks and various games embodying the group structure.

In Dienes' experimental study, the perceptual variablity principle is virtually taken for granted, in that mathematical structures are presented to all the experimental groups in multiple embodiments. One group (group A), however, was given a considerable amount of work in one embodiment before proceeding to the next: at first, for example, they worked exclusively with wooden squares and cubes. In the work with other groups, there was an immediate mixture of embodiments.

Dienes suggests that the distinct perceptual blocks
encountered by Group A may have been the result of working in
one situation for too long: the subjects appeared to attach
importance to features quite irrelevant to the mathematical
structures they represented. On the other hand, variation
of the embodiments does not necessarily lead to the generation
of an abstraction: young children in particular may not be
able to think in terms of some invariant structure, and may
look for some over-all perceptual similarity - what Dienes
calls 'pictorial communality' - rather than common abstract
properties. This does not necessarily invalidate the
principles; it suggests that teachers need to be aware of
whether the experiences are in fact working for the pupils in
the desired way.

Two studies have been undertaken which compare a number
of different approaches to the teaching of decimal multiplication and of directed numbers, respectively.

Bell and Beeby (1979) compare the effectiveness of two
approaches to decimal multiplication:

(a) an approach which provides justified rules and procedures
    for carrying out the calculations (Rules Presentation);

(b) an approach in which a particular embodiment of decimals
    (using squared paper) is used and explored in depth
    (Concrete Embodiment Presentation).

In this study, the Chelsea CSMS test on place value and
decimals, with additional items involving extended calculations, was used as a pre-test, post-test and delayed post-test
given three months later: over the whole period there was no
significant difference between the gains of the two groups.
The embodiment used here approaches the computation by a route
which may be too involved for many pupils; whereas the rules
group could perform calculations relatively easily, at least
while the required rules were fresh in the memory. Thus, for
short-term mastery of calculations, use of carefully selected
rules may be appropriate; this conclusion, however, is based
on observation of pupils' progress during the teaching and
does not show clearly in the test results. The results of
one question emphasize the need for specific teaching of the
meaning of decimal numbers, but it is not clear that the
single embodiment used is particularly appropriate for this
purpose. What does become clear in this study is the hazard
of using an embodiment to explain a principle: a problem such
as 2.8 x 1.3 concerns numbers which are already seen as
abstract entities, and to accept the representation of this
as the area of a rectangle is a difficult step. The embodiment appears to be useful as a starting-point for problem
formulation and analysis, but not as a reference point for an
abstract task.

In Shiu's (1979) comparison of approaches to teaching the addition and subtraction of directed numbers, the emphasis, at least initially, is on comparing teaching methods (systems and strategies, guided discovery, exposition) and not on the use of particular embodiments. However, it does become clear that the systematic exploration of a single embodiment backed up by practice of skills produces greater immediate learning than a greater variety of situations and tasks, and that this superiority is sustained over a period of time. It is remarked, however, that 'the guided discovery group patently enjoyed the variety of the application work'.

Bell (1976) describes some work with a spike abacus, which indicates some of the factors relevant to the choice of structured material as an aid to abstraction. Addition is being carried out in base 5: thus, 'when four rings are filling a peg and another two are to be added, we take away five, put one of them on the next peg and leave the sixth one on the present peg'. This sequence is very complicated: weaker pupils were unable to cope with it, particularly in respect to the four discarded rings which 'ought to have a place somewhere'. The preoccupation with manipulation does not lead to understanding of place value - rather the reverse.

Many descriptive articles have been written in which particular embodiments are suggested as approaches to certain topics, often with some comments on the writer's experience of this teaching approach. (See, for example, Malpas, 1975, Bartolini, 1976, on directed numbers.) Often these recommend a specific method, but Malpas' article is interesting in that he describes his use of the model as a 'failure'. Having introduced addition of directed numbers as 'shifts' on a number-line, he then proceeded to attempt subtraction using the same model. A plausible explanation for the lack of success of this, however, lies not so much in the limitations of the embodiment as in the introduction of an inverse process when the original operation has not yet become thoroughly familiar. This is consistent with Collis' (1974) discussion of the understanding of inverse at this stage.

In a study by Fennema (1972), selected children (about seven years old) were taught a previously unlearned concept (multiplication defined as the union of a number of disjoint sets of equal size), either by a 'concrete' or 'symbolic' treatment. Cuisenaire rods were used for the concrete treatment; 3,2→6 was modelled as: 3 two-rods end to end are equivalent in length to a six-rod. In the symbolic treatment, 3,2→6 was written 2 + 2 + 2 = 6. The task being taught was that of solving problems of the type 5,2 = □ by one or other of the methods. Results of a recall test containing problems included in the teaching, but including the form 5, □ = 10, showed that the methods were equally effective. The problems taught and tested in the Recall tests all had products 10. The Transfer tests consisted of problems with products between

11 and 16, inclusive. For the immediate tests, pupils had
the same materials as for the learning; for the delayed tests,
all children had counters available.When transfer or extension
of the principle was tested, children who used the symbolic
model performed at a higher level than those who used the
concrete model.   Quite clearly, both groups had had earlier
concrete experience, and were now able to consider the
addition of small, equal numbers without needing physical
embodiments, and the study appears to indicate that concrete
models are not always necessary or helpful when the concept
embodied is already possessed in an abstract form.

There are a number of theses which investigate the effect
of the *number* of embodiments, testing Dienes's perceptual
variability principle, mostly without significant results.
Thus both Gau (1973) and Beardslee (1973) worked with groups
of pupils using one, two or three different models for
representing and manipulating fractions, and discovered no
significant differences in their subsequent ability to work
with fractions in the usual symbolic notation.   On the other
hand, one study, by Wheeler (1972), does identify an effect.
Wheeler introduced groups of children to methods for adding
and subtracting two-digit numbers, which he explained in
terms of one, two, three or four embodiments; these were the
abacus, bundling sticks, a place-value chart and multibase
arithmetic blocks.   The number of embodiments did *not* have a
significant effect on the learning of the procedure for
handling two-digit numbers, but those children who had used
three or more different embodiments were significantly more
proficient in extending the method to three-digit numbers.
Indeed, there is a significant positive correlation between
the number of embodiments used for the two digits and
achievement on a multi-digit test.   This is, on the face of
it, a good example of how working with several embodiments
can help abstraction and transfer of the principle.

A large-scale study by Biggs (1967) of arithmetic in
primary schools contrasts 'uni-model' (Cuisenaire) and 'multi-
model' (Dienes) teaching methods (described as 'structural')
with 'traditional' methods (direct teaching of computational
processes) on the one hand, and 'motivational' methods (not
using structural materials, but making use of activities and
'real-life problems') on the other.   The study is complicated
by the investigation of the interaction of the method effects
with intelligence and sex differences.   Uni-model methods
proved more effective than traditional methods in the 'remedial' situation but not significantly different with average
children: on the other hand, multi-model methods produced
higher test scores and more favourable emotional reactions
amongst all groups of children.   Girls appeared more likely
to find initial difficulty and high anxiety with multi-model
methods, but after about two years, this disappeared and high
scores were achieved in relation to a traditional control
group.

In summary, the general conclusion of this research is that constructive activities in an embodiment provide a useful starting-point for the development of analytical thought leading to an abstraction; but where a task is already seen in abstract terms embodiments appear less helpful. Care must be taken to choose an embodiment which can develop conceptual understanding and does not become just an elaborate, procedural device for a particular category of task: the uses of squared paper for decimal multiplication and of the spike abacus described earlier tend towards the latter situation. As far as the *number* of embodiments is concerned, restriction to a single embodiment does not appear to have an adverse effect on the performance of particular, closely-related tasks, and indeed may be better for certain specific objectives than use of a variety of models. However, experience of several embodiments can improve the ability to extend and transfer understanding to other situations. This may be explained in terms of both the elimination of incidental features of particular embodiments which are irrelevant to the mathematics, and the assimilation of different facets of the conceptual structure which a new embodiment may highlight.

The discussion so far has concerned materials specially designed to embody a chosen mathematical structure, such as the base blocks embodying the place-value system of notation. But structures are also embodied in familiar environmental situations, for example, directed numbers in monetary transactions, fractions and decimals in rulers, products in dining tables each seating the same number of children. Evidence presented in Chapter 2 shows that structures embedded in familiar situations are much more easily handled. Szeminska (1965) also observed that reasoning ability was much better in familiar, real-life contexts than in material learnt only in school. However, it may be that this very familiarity means that the structural properties are harder to abstract than from special material. This question needs further research. It seems clear both that familiar situations should be more fully exploited, and also that such concrete embodiments as are used should continue in use long enough to become thoroughly familiar and their properties intimately known, and that they should continue to be available as 'fallbacks' as long as necessary; they should not simply be regarded as *ad hoc* visual aids by which a principle can be established once for all. (See Vergnaud, 1979.)

8. THE DEEP-END PRINCIPLE

Dienes and Jeeves (1963, 1970) conducted a series of experiments in which subjects endeavoured to learn a binary operation, in which colours combined according to certain group structures. Coloured discs were displayed and a machine or the experimenter responded to each displayed pair

of discs by showing a third disc, this being the result of
'combining' the first two. The subjects made a succession of
such trials and attempted to learn to predict the combinations. Effectively they had to learn the group combination
table, or, preferably, laws which would generate it, such as
'the red keeps everything the same', or 'two different
colours always give the third'. Subjects were required to
learn a number of different structures in this way, in
different orders, and the main point of interest was which
were the best orders. From this emerged the 'deep-end'
principle, since it appeared that, at least for children if
not for adults, total learning was greater when the more
complex structures were learned first, for example the
4-element group before the 2-element group, and 5 before 3.
It seemed that knowing the simpler rules which governed the
smaller structures led subjects to direct their efforts to
making these fit the more complex structures which caused
blockages, whereas seeing a simpler version of a more complex
rule was easy.

Somewhat similar conclusions were drawn by Markova
(1969), who studied how two fairly able groups of 15-year-
olds played games of Nim of varying degrees of complexity,
either in increasing or decreasing order. From children's
verbalizations, typically, imperatives such as 'take an odd
number of chips in each draw', hypotheses such as 'If I take
an odd number of chips in each draw I shall win' were
inferred. Two kinds of hypothesis were identified (H_1 and
H_2) and children were classified according to which type
they used most over the four games. Occasionally changeovers
occurred from one game to the next (6 and 7 for Groups I and
II respectively). Examples of H_1 and H_2 are given below.

Examples of protocol

Hypothesis 1: Game G_6; female M.M. Group II, Raven
Matrices I.Q.: 121.

   (a) I will try to take one chip more than you.
   (b) Now I take a half less than you.
   (c) Now I will take one chip less than you.
   (d) Now I will always take one chip.
   (e) I will take the same number as you.
   (f) Now whenever you take the highest number of chips
       I shall take the smallest.

Hypothesis 2: Game G_6 was played: male R.H., exp. group II,

   (a) I see, when seven are left I will lose. I must try
       to have my turn when there are six left.
   (b) It is the number seven that is important. How to get
       it? I must play in such a way so that the number is
       for you.
   (c) Thirteen chips remain, it depends how many you take.

Whatever you take the seven will remain to you.

(d) What will be the situation if I take four at the beginning? You can take one to five. No, I will not take four, I will take one for you to get 13.

(e) I do not know yet if I can win every time. I understand it already but I must not make a mistake. Whatever you take at the beginning I must complete the number to such an amount that 13 will remain for you and then seven.

H_1 hypotheses are more general and comprehensive than H_2, but less effective than H_2, which are based on a more detailed and abstract analysis of the problem and which focus on strategies for solving sub-goals.

H_1 hypotheses do not seem to form a system. If a particular H_1 strategy was unsuccessful, subjects tried to arrive at another H_1 hypothesis, not on the basis of analysing the situation but intuitively only. On the other hand, sub-goal solving of the problem by means of H_2 hypotheses brings the subject to the understanding of the structure of the game.

For simple games it is advantageous for the subject to try all the possible combinations of the set 'to take the same as you', 'to take the opposite of you', etc. (i.e., H_1), rather than to attempt the goal analysis (H_2). However this is less effective for more complex games, where the set is much larger. Subjects tended to choose and stick to H_1 hypotheses if they started by playing a simple game (Group I), and H_2 if they started with a more complex game (Group II):

	H_1	H_2	
Group I (complexity increase)	37	8	(Significant at .005 level)
Group II (complexity decrease)	12	33	

Though the choice of hypothesis type did not relate strongly to children's success on the games actually played (significant at the .05 level for Group II children but non-significant for Group I), only children using H_2 were likely to have sufficient understanding of the game's structure to be able to succeed at the theoretical game.

(Only children successful in the actual games were given the theoretical game: 'Now imagine that there is a pile of 50 chips. In each draw you are allowed to take

	Theoretical game	
	Unsuccessful	Successful
H_1	11	10
H_2	0	24

(Significant at .005 level)

l to 6 chips. How would you manage to win each game if I again always start first?')

This is an interesting example of the deep-end principle, where presenting children with complex items first is more effective than giving a graded set of items of increasing complexity.

This might be regarded as a special case of evoking *cognitive conflict,* in that children are presented with a situation where they can recognize that a particular (primitive) strategy is not going to be effective, but with the advantage that a set for this strategy is not first established.

It is fairly clear that the deep-end principle will not work for a given problem for children below a certain level of ability; the question then arises, does one allow these children to develop a more primitive strategy (in the belief that this is better than nothing), even though this may later produce resistance to a more advanced strategy, or does one postpone the learning of that structure until it can be approached more broadly?

This question relates to that of advance organizers and that of orientation induced by the type of task (see Computer-Based Learning, below).

9. COMPUTER-BASED LEARNING RESEARCH

The computer makes it possible to provide an environment which responds to the learner in prescribed but not inflexible ways. It also can record the learner's choices and map his progress through the material. It is therefore an apt instrument for research on teaching and learning. The computer has also inspired theory-building and experiment based on the view of the human learner as an information-processor. The research to be discussed here is related to the computer in both of these ways.

Hartley (1980) describes a range of such experiments, many of them conducted by his own group at Leeds; we shall quote a selection of these. One of these aimed to show how the kind of learning achieved depended on the orientation produced by the kind of task required of the learner. Poly- technic students were presented with a text on probability and asked questions at either a syntactic level ('Underline any words with more than six letters'), or a semantic level (either paraphrase parts of the material, or provide examples of the propositions). Three post-tests were used: free- recall, cued-recall and a comprehension test. The two semantic groups were much the superior on all post-tests: in the free-recall test they recalled sentences whilst the syntax group recalled words and short phrases. The two semantic

197

groups recalled about the same amount of material, but the paraphrase group were superior on free-recall and the 'example' group on comprehension. The experiment suggests that different tasks engage and develop different cognitive structures, which results in different learning outcomes. The implications are that teaching should generally emphasize meanings, not particular verbal forms or formulae, and that asking for examples induces deeper-level processing than asking for paraphrasing.

In another experiment subjects were taught about probability using either a formal treatment (terms were defined, and formulae used to show the relations between concepts) or diagrams and everyday terms. On a cued-recall test, the formal group was significantly the better when formal cues were given and vice versa; on a problems test which used wording that appeared to favour neither group, the informal/diagrammatic group obtained significantly higher scores - largely due to the performances of students low in mathematical experience. (See also Mayer, 1978.)

Learner-controlled learning

Computer-based teaching programs can allow the learner to proceed at his own pace, and to select which parts of the teaching materials to learn first. In one experiment, a group using a program-controlled route through the material (which was regulated by individual performance levels) performed better on post-tests than a group who could access material as they wished, even though the latter gave higher ratings of satisfaction and stimulation.

Pask (1976) distinguishes between students who in free-learning situations favour what he calls 'serialist' and 'holist' learning styles (or operationalist and comprehensive). He found that when interrelated topics A, B, C, D were presented in sequence such as A, B, AB, C, ABC, D, and finally ABCD, they suited the serialists whilst sequences like AB, CD, ABCD suited the holists better. This suggests teaching styles should be matched to learning dispositions, though Pask also found that learners are not always able to make these decisions themselves.

10. REPETITION AND THE SPACING OF REVISION

The way in which memory works and the relevance of this to mathematical education has been discussed in Chapter Three. It is clear that in mathematics, as in all other areas of learning, retention is always imperfect, and a proportion of material is forgotten over a period of time. Moreover, it may be that the particular nature of mathematical learning makes the problem of forgetting more critical: neither a lost conceptual structure nor a forgotten skill can easily be regained through simple reading.

Revision is a familiar strategy for maintaining or improving retention: its effectiveness is somewhat variable, and it is worth giving some thought to the amount of repetition needed and to the appropriate timing of revision sessions. For meaningful, descriptive material in prose passages, some research points to the effectiveness of practice in the form of spaced reviews rather than massed revision at the end of a long course of reading (see, for example, Ausubel and Youssef, 1965). A study by Reynolds and Glaser (1964) investigated the effect of spaced reviews on the learning of scientific material. In this investigation, two groups, scoring equally on a pre-test, were taught elementary biology in a programmed sequence of 20 sessions, each lasting for 40 minutes. With one group, the programme was interrupted by short spaced review sessions, part way through: the other group was given massed repetition of the material after the initial programme was completed. The spaced review group achieved better retention of the material at both 10 days and 31 days after the last review of the particular topic being tested. While this is an isolated experiment, it shows clearly that the timing of revision sessions is a significant factor.

A slightly different question has been considered by Gay (1973) who looked at the effect of the timing of reviews on the retention of mathematical 'rules'. Her study is concerned with the situation in which a rule is learned in a relatively short session and then revised some time later. 'Rules' included in the work include finding the third angle of a triangle when two are given, finding the product of powers (e.g., $a \times a^2 \times a^3$) and finding a geometric mean. In her first experiment one review was given either one day, one week or two weeks after the original learning: a retention test after three weeks revealed that while all review groups retained significantly more than a no-review group, the timing of the revision was not a significant factor.

In a second experiment, Gay gave two reviews to each of three groups: these were placed one and two, one and seven, and six and seven days after the original learning and again the groups were tested three weeks after the original learning. Two reviews appear to be more effective than one: these groups retained between 75 per cent and 150 per cent more than the no-review group, compared with about 47 per cent for the one-review groups. The timing, however, is more critical here: the one and seven days pattern is the best, and one and two the worst. The suggestion here must be that review sessions should not be too close in time; but it is not clear from this just what the optimal placing might be. It may be that an early (one day) review serves to re-establish the material before too much interference has taken place and that, subsequently, spaced reviews are more successful in renewing its retrievability.

References

ANDERSON, J.E. (1964): *Psychology of Development and Personal Adjustment*. New York: Holt.

ASHLOCK, R.B. and HERMAN, W.L. (1970): *Current Research in Elementary School Mathematics*. New York: Macmillan.

ANTHONY, W.S. (1973): 'Learning to discover rules by discovery', *J Ed Psych,* 64, 325-8.

AUSUBEL, D.P. (1968): *Educational Psychology: A Cognitive View*. New York: Holt, Rinehart and Winston.

AUSUBEL, D.P. (1978): 'In defense of advance organisers: a reply to the critics', *Rev Ed Res,* 48.2, 251-7.

AUSUBEL, D.P. and YOUSSEF, M. (1965): 'The effect of spaced repetition on meaningful retention', *J Gen Psych,* 73, 147-50.

BARNES, B.R. and CLAWSON, E.W. (1975): 'Do advance organizers facilitate learning? Recommendations for further research based on an analysis of 32 studies', *Rev Ed Res,* 45, 637-59.

BARTOLINI, P. (1976): 'Addition and subtraction of directed numbers', *Maths Teaching,* 74, 34-5.

BASSHAM, H. (1962): 'Teacher understanding and pupil efficiency in mathematics - a study of relationship', *Arith Teacher,* 9.

BEARDSLEE, E.C. (1973): 'Toward a theory of sequencing: study 1-7: an exploration of the effect of instructional sequences involving enactive and iconic embodiments on the ability to generalise', Pennsylvania State University. *Diss Abs Int,* 33A, 6721.

BEILIN, H. (1965): 'Learning and operational convergence in logical thought development', *J Exp Child Psych,* 2, 317-39.

BELL, A.W. (1976): *The Learning of General Mathematical Strategies*. Shell Centre for Mathematical Education, University of Nottingham.

BELL, A.W. and BEEBY, T. (1979): *Two Approaches to the Teaching of Decimal Multiplication*. Shell Centre for Mathematical Education, University of Nottingham.

BENNETT, N. (1976): *Teaching Styles and Pupil Progress*. London: Open Books.

BIGGS, J.B. (1967): *Mathematics and the Conditions of Learning*. Slough: NFER.

BRAINERD, C.J. (1978): 'Learning research and Piagetian theory'. In: BRAINERD, S.: *Alternatives to Piaget*. Academic Press.

BROWNELL, W.A. (1945): 'When is arithmetic meaningful?' *J Ed Res*, 38, 481-98.

BROWNELL, W.A. and CARPER, D.V. (1943): *Learning the Multiplication Combinations*. Duke University Res Stud Ed (7), Durham, NC: Duke University Press. p.177.

BROWNELL, W.A. and CHAZAL, C.B. (1935): 'The effects of premature drill in third-grade arithmetic', *J Ed Res*, 29, 17-28.

BROWNELL, W.A. and MOSER, H.E. (1949): *Meaningful Versus Mechanical Learning: a Study in Grade II Subtraction*. Duke University Res Stud Ed (8), Durham, NC: Duke University Press. p.207.

CASE, R. (1975): 'Gearing the demands of instruction to the developmental capacities of the learner', *Rev Ed Res*, 45.

CASE, R. (1978): A developmentally based theory and technology of instruction. *Rev Ed Res*, 48.

COLLIS, K.F. (1975): *Cognitive Development and Mathematics Learning*. Chelsea College, London.

CORMAN, B.R. (1957): The effect of varying amounts and kinds of information as guidance in problem solving. *Psych Monog*, 71, p.2.

CORY, R. (1981): Evaluation of individualised learning materials. M.Ed. thesis, Bath.

COSGROVE, G.E. (1957): The effect on sixth-grade pupils' skill in compound subtraction when they experience a new procedure for performing this skill. Doctoral dissertation, Boston University School of Education.

CRANNELL, C.W. (1956): 'Transfer in problem solution as related to the type of training', *J Gen Psych*, 54, 3-14.

CRAWFORD, D.H. (1975): *The Fife Mathematics Project*. London: O.U.P.

DIENES, Z.P. (1959): *Concept Formation and Personality*. Leicester University Press.

DIENES, Z.P. (1960): *Building Up Mathematics*. London: Hutchinson.

DIENES, A.P. and JEEVES, M.A. (1963): *Thinking in Structures*. London: Routledge and Kegan Paul.

DIENES, Z.P. and JEEVES, M.A. (1970): *The Effect of Structural Relations on Transfer.* London: Routledge and Kegan Paul.

DONALDSON, M. (1978): *Children's Minds.* Fontana.

EULL, W. and SILVERMAN, S. (1970): 'Learning Set as a technique for inducing conservation', *Eastern Psych Ass.* Atlantic City, New Jersey.

FENNEMA, E.H. (1972): 'The relative effectiveness of a symbolic and a concrete model in learning a selected mathematical principle', *J Res Math Ed,* 3, 233-8.

GAU, G.E. (1973): 'Toward a theory of sequencing: study 1-6: an exploration of the effect of instructional sequences involving enactive and iconic embodiments on the attainment of concepts embodied symbolically. *Diss Abs Int,* 33A, 6728.

GAY, L. (1973): 'Temporal position of reviews and its effect on the retention of mathematical rules', *J Ed Psych,* 64, 171-82.

GELMAN, R. (1969): 'Conservation acquisition: a problem of learning to attend to relevant attributes', *J Exp Child Psych,* 7, 167-87.

GILES, G. (1972): *The Fife Mathematics Project Report.* Stirling University.

GILES, G. (1978): *School Mathematics Under Examination (2).* Stirling University.

GOLD, A.P. (1978): Cumulative learning versus cognitive development. Ph.D. thesis, University of California, Berkeley.

HARTLEY, J.R. (1980): Using the computer to study and assist the learning of mathematics. Paper presented to the BSPLM, Shell Centre, University of Nottingham, January.

HENDRIX, G. (1947): 'A new clue to transfer of training', *Elementary Sch J,* 48, 197-208.

HIRSCH, C.R. (1977): 'The effects of guided discovery and individualised instructional packages on initial learning, transfer and retention in second-year algebra', *J Res Math Ed,* November, 359-68.

HOWARD, C.F. (1950): 'Three methods of teaching arithmetic', *Calif J Ed Res,* 1, 3-7.

HYTCH, J.D. and TIDMARSH, D.E. (1969): 'Mathematics, streamed and non-streamed: a practical approach at the secondary stage', *Maths Teaching,* 48.

INHELDER, B., SINCLAIR, H. and BORET, J. (1974): *Learning and the Development of Cognition.* London: Routledge and Kegan Paul.

KATONA, G. (1940): *Organising and Memorising.* New York: Columbia University Press.

LAND, F.W. (1970): *Technological Aids in the Teaching of Mathematics.* Aspects of Educational Technology IV, London: Pitman.

LAND, F.W. and BISHOP, A.J. (1967-70): *Annual Report of the Mathematics Teaching Research Project.* University of Hull Institute of Education.

LANGDON, N. (1976): Smiling in the classroom. *Maths in School,* 5, 5, November.

LAWTON, J.T. (1977): 'The use of advance organizers in the learning and retention of logical operations and social studies concepts', *Am Ed Res J,* 14, 1, 25-43.

LAWTON, J.T. and WANSKA, S.K. (1977): 'Advance organizers as a teaching strategy: a reply to Barnes and Clawson', *Rev Ed Res,* 47, 233-44.

LAWTON, J.T. and WANSKA, S.K. (1979): 'The effects of different types of advance organizers on classification learning', *Am Ed Res J,* 16, 3, 223-39.

LESH, R.A. (1976): 'An interpretation of advanced organizers (and other titles), *J Res Math Ed,* 7, 69-91.

LYDA, W.J. and MORSE, E.C. (1963): 'Attitudes, teaching methods, and arithmetic achievement', *Arith Teacher,* 10, 136-38.

MALPAS, A.J. (1975): 'Subtraction of negative numbers in the second year', *Maths in School,* July.

MARKOVA, I. (1969): 'Hypothesis formation and problem complexity'. *J Exp Psych,* 21, 29-38.

MAYER, R.E. (1978): 'Effects of meaningfulness on the representation of knowledge and the process of inference for mathematical problem solving', In: REVLIN, R. and MAYER, R.E. (Eds): *Human Learning.* Winston.

MILLER, R.L. (1979): 'Individualised instruction in mathematics: a review of research', *Maths Teacher.*

MORGAN, J. (1977): *School Mathematics Under Examination (1): Affective Consequences for the Learning and Teaching of Mathematics of an Individualised Learning Programme.* Stirling University.

MURRAY, F.B. (1972): 'Acquisition of conservation through social interaction', *Dev Psych*, 6, 1-6.

PASK, G. (1976): 'Conversational techniques in the study and practice of education', *B J Ed Psych*, 46, 12-25.

PAIGE, G. and SIMON, H.A. (1966): 'Cognitive processes in solving algebra word problems'. In: KLEINMUNTZ, B. (Ed): *Problem Solving.* Wiley.

REYNOLDS, J.H. and GLASER, R. (1964): 'Effects of repetition and spaced review upon retention of a complex learning task', *J Ed Psych*, 55, 297-308.

ROGERSON, A. (1979): 'Individualised learning', *Int J Math Ed Sci Tech*, 10, 2, 199-204.

ROMISZOWSKI, A.J. (1979): 'What's happening to individualised mathematics', *Programmed Learning and Ed Technol*, 16, 2, 146-50.

ROUGHEAD, W.G. and SCANDURA, J.M. (1968): 'What is learned in mathematical discovery?' *J Ed Psych.*

SCANDURA, J.M. *et al.* (1969): 'An unexpected relationship between failure and subsequent mathematics learning', *Ed Stud Math*, 1, 247-51.

SCHOEN, H.L. (1976): 'Self-paced mathematics instruction: how effective has it been?' *Arith Teacher.*

SHEPPARD, J.L. (1974): 'Compensation and combinatorial systems in the acquisition and generalisation of conservation' *Child Dev*, 45, 717-30.

SHIU, C.M. (1979): *Teaching the Addition and Subtraction of Directed Numbers: A comparative Study.* Shell Centre for Mathematical Education, University of Nottingham.

SHULMAN, L. (1970): 'Psychology and mathematics education'. In: *Mathematics Education*, 69th Yearbook of the National Society for the Study of Education.

SILVER, E. (1978): Students' perceptions of relatedness in word problems. Paper presented at AERA.

SKEMP, R.R. (1971): *The Psychology of Learning Mathematics.* Penguin.

SWENSON, E.J. (1949): 'Organisation and generalisation as factors in learning, transfer and retroactive inhibition'. In: *Learning Theory in School Situations,* Studies in Education, No.2. Minnesota: University of Minnesota Press.

SZEMINSKA, A. (1965): 'The evolution of thought: some applications of research findings to educational practice', *Monog Soc Res Child Dev,* 30.

TREMBATH, R.J. and WHITE, R.T. (1979): 'Mastery achievement of intellectual skills', *J Exp Ed,* 47, 247-52.

TROWN, E.A. (1970): 'Some evidence on the interaction between teaching strategy and personality', *B J Ed Psych,* 40, 209-11.

TROWN, E.A. and LEITH, G.O.M. (1975): 'Decision rules for teaching strategies in primary schools: personality-treatment interactions', *B J Ed Psych,* 45, 130-40.

VERGNAUD, G. *et al.* (1980): *Didactics and Acquisition of Multiplicative Structures in Secondary Schools.* Cognitive Development Research in Science and Mathematics, University of Leeds.

WASON, P.C. and SHAPIRO. (1971): 'Natural and contrived experiences in a reasoning problem', *Quarterly J Exp Psych,* 23.

WHEELER, L.E. (1972): The relationship of multiple embodiments of the regrouping concept to children's performance in solving multi-digit addition and subtraction examples. Ph.D. thesis, Indiana University.

WORTHERN, B.R. (1968): 'Discovery and expository task presentation in elementary mathematics', *J Ed Psych,* 59, 1-13.

YATES, J. (1978): *Four Mathematical Classrooms: an Enquiry into Teaching Method.* University of Southampton.

Chapter Eight
Process Aspects of Mathematics

1. INTRODUCTION

The style and atmosphere of the activity in the mathematics classroom is a strong component of what is learned by the pupils: whether they see mathematics as a field of inquiry, or a deductive system, or a set of methods to be learnt from the teacher - or a combination of these. This complex of general concepts of mathematics, with their associated strategies to be learnt by the pupils - how to practise skills, how to solve problems, how to recognize and to construct proofs, how to generalize - will be treated in this chapter under the heading of Process Aspects of Mathematics.

It is generally accepted that being mathematically competent involves more than knowing a range of particular concepts, theorems and skills. This additional expertise is described by some workers as ability for mathematical activity, ability to act like a mathematician, or competence in problem-solving with the use of particular heuristic strategies. This area of activity may be considered under the three main headings of:

1) Investigation and problem solving;
2) Proof;

3) Representation, generalization and abstraction.

The diagram below illustrates some of the relations among these notions.

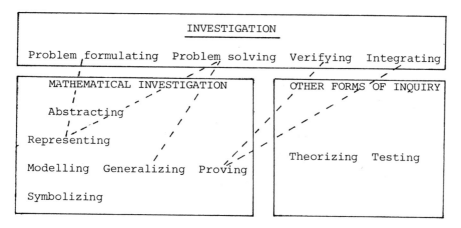

Components of the process of investigation.

Here the term 'investigation' is used in an attempt to embrace the whole variety of means of acquiring knowledge. This requires the important addition to problem-solving of *problem formulating*, and also of the phase of *integrating,* that is of incorporating the new item of knowledge in the body of previously known knowledge, and of reviewing the newly enlarged collection of knowledge as a whole. These four phases are associated loosely with the usually distinguished features of mathematical investigation, for example the problem formulating phase is one which draws most strongly on the mathematical activities of representation and modelling, but these features are also involved in the problem-solving phase. Similarly, mathematical proof is involved both in the verifying phase in supporting a conjecture or a sketch proof by an articulated deductive argument, but proof in the sense of the construction of a deductive *system* plays an important part in integrating the new knowledge with the old. The lower part of the figure notes different modes of mental operation which occur in all phases of the process of investigation.

Stages of development in these aspects of mathematics and related teaching experiments are discussed in this chapter.

2. PROBLEM SOLVING

Most of the current research on problem-solving being carried out in relation to mathematical education has its origins in Polya's work; some more recent work has used Newell and Simon's (1972) information-processing approach. The aim of the Polya-based studies has generally been to see to what extent it is possible for students to learn and apply the set of heuristics based on 'understanding the problem, devising a plan, carrying out the plan and looking back' (Polya, 1945). The most recent and extensive research of this type is that of Kantowski (1979). She and her colleagues chose the following heuristics for study:

1. Heuristic processes related to *planning*

    1–1 Posits goal-oriented plan
    1–2 Posits plan for intermediate goal
    1–3 Uses trial and error
    1–4 Searches for pattern
    1–5 Sets up table or matrix

2. Heuristic processes related to *memory for similar problems*
   (during solution)

    2–1 Recalls similar problems
    2–2 Refers to method used in similar problem
    2–3 Refers to fact(s) found in solution of similar problem
    2–4 Refers to structure of similar problem

3. Heuristic processes related to *looking back*
   (after at least one solution is found)

    3–1 Checks solution
        3–1a checks computation
        3–1b checks steps in reasoning
        3–1c checks to see that conditions of problem are fulfilled
    3–2 Refers to related problem encountered previously
    3–3 Poses new related problem
    3–4 Attempts to find another, different solution
    3–5 Attempts to simplify solution

Seventeen 14-17-year-old students who had done well in normal mathematics courses were grouped into a class and were taught a course of 22 lessons based on the discussion and use of Polya's heuristic scheme.

Problem Set A–2

A 2–1. Abner Prugg has three sons. The product of the sons' ages is 200.
*   The oldest son is twice the age of the second oldest son. Find
       the ages of all three sons.

A 2–2. Point P is taken in the interior of a square whose side has length
*   8 cm. P is equally distant from two consecutive vertices of the
       square and from the side opposite these vertices. Find the
       distance from P to one of the vertices.

A 2–3. The diagram shows a triangular
*   grid that is three units
       on each side.
       How many triangles one unit
       on each side are there in
       a grid that is 20 units on
       a side?

A 2–4. Nowadays, no one writes a novel without dedicating it to someone.
       Each of five novelists whose works have recently appeared has
       dedicated his book to the only daughter of one of the others. No
       daughter is the recipient of more than one dedication.

       Mr. Scarlett's novel is dedicated to Mr. White's daughter, Frances.
       Mr. White's novel is dedicated to Mary; Mr. Lemmon's to Annie;
       Mr. Greene's to Joan.

       Mary is the daughter of the novelist who dedicated his book to
       Mr. Black's daughter.

       Kate is Mr. Lemmon's daughter.

       Who is the father of her friend, Anne?

There were also problem-solving sessions, interspersed through the course in which students solved problems from a set including four types, number generalization, number theory (factors), logical and geometrical. Examples of each are given on the previous page.

The results reported include the following: number problems of type A2-1 evoked very few goal-oriented plans; 'guess and check' was the most common approach; only three of the 17 students went on to seek further solutions after finding one, 'indicating very poor use of "looking back"'. In several of the geometry problems goal-oriented planning was evident, in the logic problems most students 'looked back' by checking that the found solution satisfied the conditions; in these problems, using a table to display all possible combinations was a heuristic taught, and eventually adopted. The specific heuristics for the number generalization problems of 'start with a simple case', 'set up a table', 'find a pattern and generalize' were readily accepted and used. Goal-oriented planning of some sort was found in a high percentage of correct solutions, though this might consist simply of 'trial and error'. On reflection it was thought that the looking back process had not always been made an integral part of the problem-solving activities presented in the instructional sessions. Difficulties were noted in seeing how the *general* heuristics applied to the particular problems, whereas the heuristics specific to each problem type were more readily adopted. It was also observed that students often did not begin to use a heuristic until some time after it had been introduced. The style of the instructional sessions appears to have begun as demonstrations by the teacher of problem solutions, with attention drawn to heuristics. 'Students took a more active role as the sessions progressed, but were very hesitant to participate initially.'

The implications to be drawn from the report of this research would appear to be that there was little learning of the *general* heuristics in a way which enabled them to be transferred to new types of problem; but there was substantial learning of heuristics specific to each problem type. (Quantitative data are not presented in the report.) Together with the results of the other studies to be presented here, this suggests that more intensive instruction in each heuristic is necessary, using a collection of problems or part-problems specially chosen to practise each named heuristic until this is fully mastered, after which problems could begin to be mixed, requiring the *selection* of an appropriate heuristic. In other words, the heuristics should be taught as concepts, and named, with the objective that, for example, the term 'making a plan' evokes in the student's mind a range of well-defined plans which he has used for different problems. A second implication might be that the level of generality of the heuristics being taught needs to be adapted

to the students. After acquiring a range of specific or
middle-level heuristics for different classes of problem,
students might be able to learn and use the more general ones.

The current work of a British curriculum development and
research project in the Polya tradition is reported by Burton
(1980). An elaboration of the Polya scheme has been made,
defining nineteen aspects of procedure, under the headings
Entry, Attack and Extension, and ten aspects of skill, under
Representation and Information Analysis. A large number of
problems, classified according to the procedures and skills
they are likely to require, has been prepared and used in
12 classes in the 9-13 range in three regions of Britain.
The problems were used for twelve weekly sessions of one hour
each, and pupils' performance on a problem of standard
structure, but in two different contexts, were measured
before and after the term's teaching.

Some earlier studies

Several researches exist which bear on the learning of general
strategies. The first two to be quoted concern general
heuristic strategies and the remainder mathematical strategies, except for one which is about an aspect of scientific
method.

Covington and Crutchfield (1965) developed a General
Problem Solving Programme based on strategies such as
planning one's attack, searching for uncommon ideas, transforming the problem and using analogies. The setting is not
mathematical, but consists of stories of how two children
solve a number of puzzles and mysteries with the help of
their uncle, a high school teacher and part-time detective.
Groups of 10- and 11-year-old pupils studied this programme
with their teachers and were tested for problem-solving
ability, creative thinking and attitude. They showed considerable gains on all of these, and the gain in problem-
solving ability was still significant five months later.
Other workers followed up this work though with less successful results (Kilpatrick, 1969).

Lucas (1974) studied the effect of heuristically-
oriented teaching of a university calculus course on the
students' ability to solve problems. An experimental group
and a control group each contained about 15 students; both
groups received 'enquiry-style' instruction except that with
the experimental group, the same style was adopted during
problem-solving sessions, and 'the problems were discussed
more thoroughly for the sake of problem solving', whereas
the control group had 'an expository treatment of problem
solution'. Also, the experimental group received 12 papers
outlining and demonstrating various heuristic strategies, and
their problem solutions were graded to reward heuristic usage.

This programme lasted for eight weeks. Pre- and post-test interviews were administered: during these, each subject talked through the solution of seven problems. Significant differences on the post-test, favouring the experimental group, were found on total score for the problems, on plan and approach, and on the strategies of using a well-chosen mnemonic notation, using the method or result of a related problem and of separating and summarizing data. There was no difference in the frequency of use of diagrams, but there was a lower incidence of incorrect diagrams in the experimental group. Among the results, the first was significant at the .005 level, the others at levels of .025 or .05, which suggests that the simple, well-defined strategy of choosing a mnemonic notation is more susceptible to teaching than the others.

Brian (1966) analysed the mathematical process into (1) constructing mathematical models, (2) conjecturing, (3) settling conjectures as true or false and (4) using known or given axioms, theorems or algorithms on problems where they clearly apply. A short course (about two weeks) designed to help students acquire these processes was given to a group of 17 college students. This resulted in a significant improvement on the third process, the settling of conjectures, but not on the others. The fourth process is described by Brian as the primary aim of most present mathematics teaching. In terms of the strategies we have defined in this chapter, Brian's first process is formulating questions and making representations, and his second and third processes appear as higher and lower levels of generalization. Thus Brian's result suggests that testing generalizations may be easier to learn than making generalizations, or formulating questions, if we may assume equal emphasis on the different processes in the teaching he provided.

Wills (1967) constructed a programmed unit to teach the following problem solving procedure: (a) a difficult problem is given, requiring a certain generalization, (b) similar, but simpler problems are presented, (c) the results of these are tabulated so as to reveal a pattern, (d) the generalization suggested by the pattern is applied to the initial problem, and the result checked. The subject matter was recursive definitions and figurate numbers; the age of the students is not stated, nor is the length of the instructional period. The pre- and post-test comprised 60 problems on a wide variety of topics which could be solved by steps (b) and (c), that is, by the generation and tabulation of examples, from which a generalization can be made and applied to the problem. 561 students took part, in three groups; the first used the programmed unit only, the second also had back-up instruction from teachers, while the third was a control. Both of the first two groups made highly significant gains compared with the control, and there was no significant difference between these two. In terms of the set of strategies discussed in this chapter Wills' experiment shows successful teaching of one well-defined strategy for generalization, including the generation of relevant examples and the making of generalizations.

Wilson (1967) attempted to improve performance on theorem-proving tasks by either (a) task-specific heuristics, or (b) identifying data and conclusion and seeking to make a connection, or (c) planning a solution in general terms. The task-specific heuristics did not improve performance, even on the tasks at which they were directed; of the general heuristics, planning a solution was sucessful in only a few of the tasks. Thus this attempt to teach strategies for theorem-proving tasks was largely unsuccessful.

Post (1967) had ten classes of 12-year-olds given a six-week period of instruction and practice in the processes of problem solving, but obtained no significant differences in comparison with a control group.

Lawson and Wollman (1975) trained classes of fifth and seventh grade (10 to 11 and 12 to 13-year-olds) in controlling variables, in the Piagetian task with bending rods. Transfer was investigated (a) to another task involving controlling variables (the pendulum), (b) to one (the beam balance) involving a different aspect of formal reasoning, in this case proportional reasoning, and (c) to other tasks, e.g., Peel's passages for showing imaginative judgments. Transfer was obtained to the task (a) involving the same strategy, but not to (b) or (c). The training consisted of evoking the subjects' intuitive judgments about 'fair tests', of clarifying and exposing their judgments to them, supplying verbal forms focusing the experience, e.g., fair test, variables, all factors the same except the one being tested, and getting the pupils to describe their actions and the rationale for them.

To summarize the results of these researches, it is clear that it is possible to teach at least some of the strategies which comprise the mathematical process, but that some are easier to teach than others. The *testing* of conjectures (Brian), the adoption of a mnemonic notation (Lucas), the finding of a generalization by generating and tabulating examples (Wills) and the learning of one strategy for scientific experiment (Lawson and Wollman) are shown here as the most susceptible to teaching. Wilson's result may perhaps be attributed to a less careful identification of the strategies actually required for his tasks. But, as in all teaching experiments, fully consistent results cannot be expected because of the difficulty of specifying the teaching in sufficient detail to identify the significant aspects.

Most of the studies on problem-solving consist of training over a fairly limited period followed by an immediate post-test. More representative of the educational situation is a study by Scott (1977). This shows a significant positive effect on achievement in geometry and algebra at the age of 16, related to the experience of at least one year of an inquiry programme at the age of 11 or 12. This inquiry training was with one of two specific teachers and the method consisted of presenting the students with an event requiring explanation, such as the larger of two blocks of wood floating in liquid while a smaller-sized piece sinks to the bottom. The student's task was to ask the teacher questions that were answerable by 'yes' or 'no' until he felt that he could correctly explain why everything in the experiment happened the way it did. Each problem-solving activity was followed by a reflective strategy session during which the students analysed the questions used earlier and recorded and categorized them according to their information-gathering value. Techniques suggested in these strategy sessions included thinking of a start, middle and end to an experiment, getting all the facts, asking precise questions. The inquiry sessions were a group activity in which all students were aware of the questions being asked, and one of their number himself built up a blackboard display of the information being gained. Previous studies (Scott, 1973) had shown that such inquiry training had a significant positive effect on the analytical aspects of cognitive style and that these results persisted at least up to the age of 16, and the present study extends this result by showing the positive effect also on mathematics achievement. In answer to a questionnaire, the students attributed their greater success in geometry to the fact that the inquiry training had developed logical thought and had helped them to reason 'behind the facts', and that it had helped them in problem-solving and proof. Their success in algebra they attributed partly to the latter and partly to the contribution of the training to helping them to develop a strategy for inquiry.

The success of this work shows that powerful general heuristic schemes can be made meaningful, in terms of experiences accessible to 11-year-olds, sufficiently broadly to transfer spontaneously to other subject matter. The significant features may be (1) long-term, intensive work with a well-defined, explicit aim, (2) teaching by particular individuals, (3) reflective discussions of method, following each episode.

3. REAL PROBLEM-SOLVING

The challenges exploited by the USMES project (*Unified Science and Mathematics in Elementary Schools*) (Lomon et al., 1975) for developing the abilities of children aged six to eleven to solve 'real' problems are mostly of the practical kind. For example, one called 'Getting There' poses both the problems of helping pupils to find their way round the school building (for example by suitable provision of notices and labels) and of devising the best way to shops, playgrounds, and so on, from the school building. Another, entitled 'Classroom Management', challenges the students to devise ways of helping to keep the classroom neat and running smoothly. This project has a number of aims, one of which is to provide a real context for the development of skills of language and mathematics, but beyond this, the development of actual problem-solving strategies on the part of the children demands of the teacher that she ensures that it is they who take the decisions and plan the activity and do not simply act as assistants in carrying out *her* plan.

Evaluations of USMES have included studies of (1) the acquisition of general problem-solving strategies, (2) the effects on basic skills in language and mathematics. For the first of these a small but genuinely real problem, involving choice of variables, observation and decision-making was given to one or more subjects at a time, and they had up to half an hour in which to offer a solution. This Notebook Problem consisted of giving the student three notebooks with differing numbers and sizes of pages, quality of paper, and price, and asking him to recommend the best for purchase in quantity by the school. He has pens, pencils, erasers and a ruler available; the price is stamped on the notebook.

Shapiro (see Lomon, 1975) administered this test in 1971-72 to about 250 students in 53 USMES and non-USMES classes in a variety of grade levels and communities. On analysing the number of objectively measurable reasons given for selecting a notebook, Shapiro found that at the beginning of the year less than 35 per cent of both USMES and non-USMES classes gave any measurable reasons. At the end of the year, less than 40 per cent of the non-USMES classes gave measurable reasons, while more than 85 per cent of the USMES students gave measurable reasons. Also analysed were the number of tests suggested or performed by the subject to substantiate the reasons given. At the beginning of the year no more than 30 per cent of the students (USMES or non-USMES) in any district suggested or performed a test of their reasoning. But by the end of the year the proportion of USMES students suggesting or performing tests had risen to over 84 per cent in every one of the seven districts, while the proportion of non-USMES students was below 15 per cent in every district. The Boston University report concluded that 'it would appear that in terms of the two dependent variables studied, the USMES experience had, irrespective of units or teachers involved, a marked and positive effect on the students' problem-solving behaviour'.

For subsequent years the Playground Problem and the

Picnic Problem were developed. Both of these were more
elaborate and, as administered, less real than the Notebook
Problem. The design of a playground was often approached in
a lot that could not be used for a school playground, in
addition to the fact that students were aware that no funds
for equipment were allotted. The Picnic Problem involved a
fake map of imaginary picnic sites and no real picnic in the
offing. No significant results were obtained from these
instruments.

The effect of USMES on basic skills was evaluated by
Shann (1974). In nearly all the comparisons with control
groups which showed any differences, the trends were slightly
in favour of the USMES groups.

The USMES programme has published a number of teachers'
guides. These were initially supported by a massive
programme of in-service training, mainly involving residential
courses away from the teacher's own district. This has now
come to an end, and it appears that by no means all the
original participants have been able to maintain the work in
their schools without external support.

A recent study of real problem solving among 17-year-old
mathematics students of high ability (Treilibs, Burkhardt and
Low, 1979) explored the relative importance in over-all
modelling ability of the particular skills of generating
relevant factors, selecting the important ones, generating
relationships between factors, selecting among such relationships, and posing specific questions crucial to understanding
the problems. All these aspects, except the selection of
important factors, were positively correlated with successful
modelling. The most distinctive characteristic of the most
able modellers was a capacity to work in a 'control' mode.

In a preliminary report of a project that focuses on
real problem solving in arithmetic, Max Bell and Usiskin
(1980) argue that it is not necessary to invoke a list of
general problem-solving characteristics to explain
children's difficulties with such problems: rather the
difficulties stem from children simply not having been taught
how to select the appropriate arithmetic concepts and skills.
To help remedy this, the project aims to develop a handbook
for teachers which offers a structural classification of the
many different applications of arithmetic, for example,   ·
addition relating to joining sets and to extending quantities,
division relating to sharing and to rates, and so on.

4. PROOF

There is a small but consistent body of research evidence
concerning the understanding of proof. Williams (1978) tested
255 pupils aged 16 and over in randomly selected classes in

Edmonton, Canada, using twelve items in which two hypothetical students argued about proofs of mathematical statements related to the normal curriculum. A typical item follows:

SITUATION NINE

STATEMENT A: Let f be any factor of some number n. If n is an odd number, then f is an odd number.

Joe: 'I think that Statement A is true, Tom.'

Tom: 'Let me see. If $n = 21$, then the factors of n are 1, 3, 7 and 21. n is odd and all of its factors are odd. If $n = 45$, then the factors of n are 1, 3, 5, 9, 15 and 45. Again n is odd and all of its factors are odd. So Statement A seems to be true Joe.'

Joe: 'I also think that Statement B is true Tom.'

STATEMENT B: Let f be any factor of some number n. If f is an even number, then n is an even number.

Tom: 'Why?'

Joe: '1. Since f is a factor of n, therefore $n = f.m$. where m is some integer.
2. If f is even, then $f = 2.k,$, where k is some integer.
3. Therefore $n = f.m = (2.k).m = 2.(k.m)$.
4. Therefore 2 is a factor of n and n is even.'

Tom: 'But Joe, this only shows that Statement B is true. It doesn't show that Statement A is true.'

Joe: 'Yes it does.'

Tom: 'I don't agree. Statement B has nothing to do with Statement A.'

QUESTIONS: (a) Whose side would you be on in the above discussion?
Joe's _____ or Tom's _____?

(b) Why?

Typical responses to this item included some which failed to recognise the logical equivalence of A and B:

'I think that Joe's proof only shows statement B is true. You would need a new proof to prove statement A is true.'

There were others which recognised a connection but were either unclear or wrong about the nature of it:

'Joe is saying the same thing, but he is just using a different example.'

On the performance of the pupils, they were divided into three groups, whose characteristics Williams describes.

The lowest 20 per cent do not recognise that a counter example falsifies a generalization.

The lowest 50 per cent do not appear to see the need for proof of any proposition which they believe to be true or consider intuitively obvious.

The next 20 per cent see the need for proof of all propositions but cannot distinguish correct from incorrect proofs. In particular, they accept check of a limited number of cases as being adequate. Some of them are willing to argue from a hypothesis without needing to know that it is true.

The best 30 per cent do understand the nature of a proof as a chain of reasoning from accepted assumptions or postulates, undefined terms and previously established theorems; and that this establishes the truth of the proposition if the postulates are true. However, only about 20 per cent fully understand the significance of hypotheses and definitions, somewhat fewer understand *reductio ad absurdum,* and none in this sample recognized the equivalence of the theorem and its contrapositive (see the item quoted above).

Williams' results are supported by a previous study by Robinson (1964) of 9 to 13-year-olds, in most respects. The disagreement is that Robinson claims that 12-year-olds recognize the inadequacy for proof of the check of a limited number of cases. This is a point perhaps sensitive to direct teaching, which may account for the difference. Other evidence (see below) would support William's conclusion.

Robinson also suggests that the normal (North American) experience of Euclidean geometry has contributed little to pupils' understanding of the need for proof or to their understanding of an axiomatic system. This conclusion is based on a small subsample but would not be in conflict with other evidence. Other workers have tested the hypothesis that the teaching of the basic concepts of mathematical logic helps students to construct mathematical proofs and to judge the validity of given arguments. No significant differences between taught and control groups were found in three broadly similar experiments (Phillips 1968, Roy 1970, Mueller 1975).

Reynolds (1967) studied the proof-concepts of grammar school pupils, hypothesizing from Piagetian theory that the lower forms would be in the stage of the 'acquisition of

formal thought', and that the fifth and sixth forms would
show 'full use of formal thought'. His questions were set
in the context of Euclidean deductive geometry and precise
mathematical reasoning. For example, one question gave
sixteen examples of pairs of even numbers expressed as sums
of two primes and asked whether this showed the truth of the
conjecture for *all* even numbers. Another asked for the
completion of an argument that if two lines were *not* parallel,
the alternate angles were unequal. Reynolds concluded from
his results that even the sixth-formers showed evidence of
formal thought in this sense only occasionally, and that
the general picture was one of a steady improvement with age,
with a substantial amount of concrete thought at all levels,
including the mathematical sixths.

King (1973) reports the development and testing of a
unit of instruction on proof for able 11-year-olds. The
subject matter consisted of six theorems of the kind
suggested by the Cambridge (Mass) Conference on School
Mathematics (1963), for experiment with pupils of this age:
the first theorem was, 'If N/A and N/B, then $N/(A+B)$', while
others extended this to $A-B$, $A+B+C$, and converses. It
would appear that the *content* of these theorems was easily
understood, and that the major teaching effort required was
in expressing the proofs in concise symbolic form. This
proved possible in 17 instructional sessions with these
pupils, but one might question its value. It would justify
itself only if such modes of expression were taken up and
used elsewhere by the pupils.

Bell (1976) showed the development of proof in 11- to
17-year-old pupils as passing through stages described as
follows:

Stage 0: Non-recognition of relationship, regularity or
pattern. This includes non-expectation of regularity in the
given situation, and also the inability to work in the situation with sufficient accuracy or consistency to observe the
regularity existing in it.

Stage 1 - Pattern: Recognition of pattern or relationship,
no sense of checking its domain of applicability. Requests
for explanation or proof may be met by general restatements
of the data, lacking explanatory features.

Stage 2 - Check: Recognition that a statement of relationship applies to a *class of cases*, so that a variety of cases
must be checked, or a general argument based on an insight
covering a whole class of cases must be applied.

At Stage 2.1 the variety of cases checked is neither great
nor systematic; there is little awareness that the extrapolation is only probable, and deductive arguments consist of
fragments not firmly linked to data or conclusion or to

each other.

At stage 2.2 there is greater variety, more systematic choice of example, more cautious extrapolation, and connected though incomplete deductive chains.

Stage 3 - Proof, all cases: In this stage there is awareness of the need to deal with all cases, so if empirical methods are used it is with explicit acknowledgement of their limitations; deductive chains are complete (or recognized not to be) and apply to the whole class of cases of which the pupil is aware.

Stage 4 - Deductive system: Awareness of need for explicit statement of starting points of arguments, and of definitions used. Mistakes may occur from failure to detect circularity in arguments.

Other work showed that 11-year-old pupils are capable of making progress in the ability to generate examples to satisfy given conditions or to test a given conjecture, in expressing a relationship in general terms, either algebraically or verbally, and, to a smaller extent, in establishing a result by checking all cases where the number of cases is relatively small. The ability to explain a result by identifying the underlying structure, and to identify and connect data and conclusion, did not appear to be generally accessible at this stage. Different results were seen as separate, and connected, but not in any sense as part of a deductive hierarchy.

Other work with 15-year-old pupils has divided proof responses into those which relied entirely on empirical checks, partial or complete, and those which attempted deductive explanations, and in the latter category found a preponderance of pseudo-explanations describable as *general re-statements*; these brought together the data and conclusion into a single statement but added nothing of an explanatory nature, that is, nothing which connected aspects of the data together or related them to external general principles (Bell, 1976, 1979).

Examples of two problems from this study will be given. The problem 'Adding a Nought' shows the difficulty of giving an explanation of a well-known generalization.

ADDING A NOUGHT

If you want to multiply by ten, you can add a nought: for example, 243 x 10 = 2430.

1. Is this true for all whole numbers?

2. Explain why your answer is right.

The response below simply gives three examples of the
use of the principles, as if this proved its validity
(Stage 1). This carries no sense of explanation, nor even
of checking the principle.

① yes it is true,

② because ~~What~~ What ever Whole number you
× by 10 you just add A 0

eg ① 100146 × 10 = 1001460

② 4766429 × 10 = 47664290

③ 276428 × 10 = 2764280

That is Why I think I am right.

Next, the problem Midpoints:

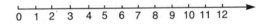

A and B can be any two whole number points on the number
line. M is the point half way between them.

1. If A is at 2 and B is at 8, at what number is M?

2. Add A's number to B's number and halve the result.
   Do you get M's number?

3. Will the rule in No. 2 work for every possible
   position of A and B on the line, including bigger
   numbers?

4. Explain why your answer is true.

This problem illustrates the difficulty in some situations
of separating data and conclusion. 'It's the average' was
a thought which linked the midpoint and (A + B)/2 and made
them hard to separate. See the response below.

If 2 numbers are added together eg 2 and 8
and then halved giving (4) The middle point
will be the same because you are finding
the distance between those two numbers.

This response and that below show another aspect of proof. It is the use of a particular case, as if it were general, in the course of a general argument. This is an important and widespread psychological phenomenon.

Because the number before M, (A) is Always the same distance away from M as B is e.g.

$$35 \quad 36 \quad 37$$
$$A \longleftrightarrow M \longrightarrow B$$

eg. 2. $14 \longleftrightarrow 17 \longleftrightarrow 20$
$\quad\quad\quad 3 \quad\quad 3$

∴ the two numbers when added together must be twice the amount of M.

Also see eg. 2. $14+5+6+7+2$ /x/x $20-3 = 17$ add this 3 to 14 and you are given 17 ∴ if you add these it equals 34 which when halved equals 17,
This works for all casses

To ascertain which of these proof concepts were most easily improved by teaching, an experiment with sixth-form pupils (aged 17) was conducted by Bell and Edmonds (Bell, 1976). The criterion tests in this case required the judgement of the validity from complete or incomplete sets of cases, examples of complete explanatory arguments, of fragmentary explanations and of general restatements of the data containing no explanatory quality. The teaching included some discussion of these points in relation to proofs written by the different pupils and passed around the class for discussion of validity. The results showed that the ability to detect an incomplete set of cases was improved, but the ability to detect a complete explanation was not significantly affected. For example, one question in the criterion test required the judgements of the validity of the following two arguments in a problem called 'Add and Take'. In this problem it is supposed that a number between 1 and 10 is added to 10 and then taken from 10, and the two results added. The question is whether the result will always be the same and why. The responses proposed are as follows:

Susan:
The result will always be 20. If you choose a number between 1 and 10 and add it to 10, then if you take the first number away from 10 it will be whatever is needed to make 20.

Yvonne:
Always 20. Whatever you add you always take it away so it cancels out. But as you add 10 and take it from 10, you get double 10 which is 20.

Have these pupils proved their answers?

Susan's: Yes/No.   Yvonne's: Yes/No.

Give your reasons:

Although it is clear to us that the second of these
arguments contains the essential points of a valid explanation while the first merely re-states the problem, this
judgement was found difficult by the pupils who tended to
demand further justification of the fact that adding and
subtracting the same number gives zero. This perhaps high-
lights the fact that in actual proof-activities judgement is
always required regarding which aspects of the argument can
be assumed as obvious and which require exposure. Further
discussion of teaching for the development of proof appears
in Bell (1976, 1978, 1979).

5.   STAGES A, B, C AND VAN HIELE'S LEVELS

In 1923 a report of the Mathematical Association, following
an even earlier Board of Education report, identified stages
A, B and C in the teaching and learning of geometry. These
are described as the experimental stage, the deductive stage
and the systematizing stage respectively, distinguished by
the kinds of verification, argument and proof which are
appropriate at each level. Thus, the role of geometry in the
curriculum was very much to develop the ideas of justification and deduction and of mathematical proof.

> *Stage A* is a 'preliminary experimental stage', based
> on practical situations, and on drawing and measuring.

> In *stage B,* the deductive stage, the pupil 'learns to
> prove theorems and riders and to write out proofs'.

> The aim of *stage C,* the systematizing stage, is 'to
> arrange the theorems in a logical sequence depending on
> a comparatively small number of axioms'.

In general, school geometry has not normally gone beyond
stage B.

More recent work on levels of thought development in
geometry has been carried out by Van Hiele (1959), who
discusses five levels of development. During the period
1960 to 1964, this work was intensively investigated by
empirical study in the Soviet Union, and is discussed by
Wirszup (1976). The Van Hiele levels are given here as
formulated by Wirszup, following Pyshkalo and Stolyar.

> *Level I* 'is characterised by the perception of geometric
> figures in their totality as entities ... judged
> according to their appearance. The pupils do not see the

parts of the figure, nor ... relationships among
components ... and among the figures themselves. The
child can memorise the names of these figures relatively
quickly, recognising the figures by their shapes alone.'

In *level II*, the pupil 'begins to discern the components
of the figures; he also establishes relationships among
these ... and between individual figures. The properties of the figures are established experimentally: they
are described, but not yet formally defined.'

(It may be noted that levels I and II are both within
the Mathematical Association's stage A, and that most of
the work described as stage A in the 1923 report is
very clearly in level II.)

Pupils who have reached *level III* 'establish relations
among the properties of a figure and among the figures
themselves. The pupils are now able to discern the
possibility of one property following from another, and
the role of definition is clarified ... The order of
logical conclusion is developed with the help of the
textbook or the teacher.' The work described as
belonging to stage B becomes accessible to pupils at this
level.

At *level IV* 'the significance of deduction as a means of
constructing and developing all geometric theory' is
recognized. The role of axioms becomes clear, and 'the
students can now see the various possibilities for
developing a theory proceeding from various premises.'

Level V 'corresponds to the modern (Hilbertian) standard
of rigour. A person at this level develops a theory
without making any concrete interpretation.

A point made by Van Hiele (1959, p.201) may clarify the
process:

  At each level, there appears in an extrinsic manner what
  was intrinsic on the preceding level. At the first level,
  the figures were in fact just as determined by their
  properties, but one who is thinking at this level is not
  conscious of these properties.

Van Dormolen (1977) refers to Van Hiele's levels of thinking
in an article about learning to understand proof. He refers
to level I as the 'ground level', and interprets this as what
Freudenthal (1973) calls 'local organisation'. 'This trapezium
is different from that one. This square has nothing to do
with that one'. In level II, Van Dormolen's 'first level of
thinking', it is discovered that certain properties of a
trapezium go for all trapezia. After working for some time
at this level, it is realised that arguments on utterly

223

dissimilar areas have elements in common. These arguments
are themselves examples leading to an understanding of what
logical organization as such is.

Thus the 'second level of thinking' is reached.

The illustrative examples given include the following:

Task: Draw an isosceles trapezium and prove that the
diagonals are equal.

Various solutions:

0. One student measures the diagonals with a ruler.
   'I find the same results, so the diagonals are equal'.

1. Another student mentally cuts out the trapezium, turns
   it about and puts it back into its hole.
   'This can be done, as one can easily see, so the
   diagonals are equal'.

2. 'An isosceles trapezium is by definition a quadrilateral
   with an axis of symmetry that does not pass through a
   vertex.

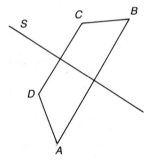

Consequently, there is a reflection S such that
$S(ABCD) = BADC$. Thus $S(AC) = BD$. With reflections a
line and its image are congruent. So the diagonals have
the same length'.

This interpretation by Van Dormolen is not entirely
faithful to Van Hiele's original scheme, but it is helpful
in understanding how the transition between levels occurs.
Thus, through classification of shapes in ground-level thinking, an understanding that all trapezia have something in
common is developed, and this makes transfer to the first
level possible. Similarly, through discussion and argument
about a variety of different situations, an understanding that
these arguments have something in common arises and this
permits the development of logical organization and deduction.

A discussion of the understanding of reflection and
rotation is given by Küchemann (1977) in the report of one of

the CSMS tests. In the approaches to reflection in particular, Küchemann identifies four stages, called global, semi-
analytic, fully-analytic and analytic-synthetic. At the
global level, the object is considered as a whole, and
reflected as a single object. In the semi-analytic approach,
part of the object is reflected first and the rest immediately
drawn in. In the fully-analytic approach, the object is
reduced to a set of points (e.g., 3 for a triangle) and these
are reflected before lines are drawn: sometimes the finished
image looks wrong. In the analytic-synthetic approach, the
two aspects are co-ordinated: image-points are located, but
the total image is assessed.

This work of Küchemann is not closely related to the
work on levels already described, since we are asking for
the performance of tasks which belong only to levels I and
II: there is no justification other than experimental and
visual verification. It is in itself worth comment, however,
that children in the 13 to 15 age range are frequently
required to work in this way; and it is remarkable that a
considerable proportion of these perform only at level I.

From the point of view of school geometry, it seems
reasonable to focus attention on levels I, II and III, while
recognizing that higher (and indeed, lower) levels exist.
The levels might be named as (I) global/descriptive, (II)
experimental/analytic and (III) deductive.

Wirszup (1976) indicates that under the old (1960)
curriculum in the Soviet Union, only 10 to 15 per cent of
pupils at the end of fifth grade (i.e., age 12) had reached
level II. It was concluded that radical and qualitative
changes in the geometry curriculum were needed; and it is now
said to be possible to begin studying 'semi-deductive'
geometry (at level III) in grade 4 (age 10-11). This,
however, depends on a detailed and planned geometry course in
the previous three years, taking pupils through level I (in
one year) and then level II.

Evidence of an awareness of the distinction between
levels II and III in particular, and that level II represents
a more advanced stage of thought, can be found in recent
developments in school mathematics in Britain.

The Scottish Mathematics Group scheme, *Modern Mathematics
for Schools* (1965-68), attempted in its first edition to make
all the work in geometry a coherent whole. Brodie (1972) says
that it is based on two axioms (although the word 'axiom' is
not used), from which the properties of various shapes can be
established. Wynne-Willson (1977) points out that this
results in extended stretches of entirely abstract work. In
the second edition (1971-74), however, the deductive work is
consistently delayed, mostly to books 6 and 7. A section on
'rotations', for example, involving deductive argument,

appears in book 4 in the first edition and book 6 in the
second. This tendency has removed some of the distinctive
flavour of the books, and brought them closer in spirit to
SMP.

In the School Mathematics Project books (A to H and
X, Y, Z), the CSE course (books A to H) treats geometry almost
entirely as an experimental science, not a deductive one.
In the few places which suggest deductive argument, the pupil
is presented with the essential argument and only asked to
follow the reasoning. This is from Book C:

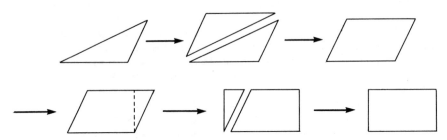

'Explain why this helps you to find the area of a
triangle'

In books X, Y, Z, which provide an O-level course,
deductive reasoning is introduced into the explanations only
very slowly. Book Y includes traditional proofs of the angle
properties of circles, and Book Z has a ten-page section on
'Proof'.

It is clear that SMP has changed its emphasis in this
respect. Quadling (1972, p.219) writes:

  The emphasis is no longer on proof for its own sake, but
  as one means of increasing our geometrical knowledge. In
  the earliest days of SMP we had thought of developing the
  whole geometrical content of the course through transformation methods; 'Can you find a proof of ... by
  transformation?' was a question frequently passed round
  the group.

However, as Wynne Willson (1977) again comments, 'in books
A - Z there are no proofs by transformation methods; only
proofs by other methods and proofs about transformations'.

It is worth correcting two possible misconceptions. This
reduction and delay in the introduction of deductive arguments
is not either (a) merely part of an over-all reduction in the
amount of geometry (SMP is at least 40 per cent geometry), or
(b) intrinsically related to the change in content (deductive
work can quite easily be based on transformations; see for
example Maxwell, 1974, or Bell, 1972-74).

226

What is apparently taking place here is a deliberate postponement of level III work until about age 14, and the restriction of this work to O-level as distinct from CSE courses.

The implications of this for teachers and curriculum constructors depend on the reason for including geometry. Wirszup's discussion of the Russian research seems to suggest that geometry can be used fairly generally and at a relatively early age as a means of developing the appreciation of deductive argument and the need for proof. However, as we have just seen, the trend in Britain appears to be against this. On the other hand, even if the development of understanding of proof is not thought to be an important feature of geometry for all pupils, the recognition of levels still has its implications. It is important to recognize higher-level tasks, on which pupils may be unable to operate unless they have made the transfer to that level; and it may be possible, by appropriate teaching, to encourage that transfer.

Usiskin (1980) reports on the early stages of a research project that intends to investigate just this question, as well as looking more generally at the usefulness of Van Hiele's theory for teaching geometry to American High School students. So far, it has attempted to specify the levels in more detail and to relate them to particular exercises in a geometry course.

6. KRUTETSKII'S INVESTIGATIONS

Krutetskii (1968, translated 1976) has produced an extensive and fascinating study of the processes used by children in solving mathematical problems. In particular he has devised a highly original series of problem types which should prove a fruitful resource for other researchers. However, Krutetskii's primary interest is to identify the special abilities (called 'mathematical abilities') of children who are mathematically highly gifted, and as such his findings give little indication of how problem-solving processes develop, or can be developed, in normal schoolchildren. As far as 'mathematical giftedness' itself is concerned, Krutetskii's arguments about how this is acquired are somewhat circumspect, in deference, perhaps, to the climate in which he worked: it is not innate but depends on certain 'inclinations'; on the other hand it seems to depend on special inborn functional characteristics of the brain, so that not everyone can become a talented mathematician ...

As a result of his investigations, Krutetskii presents the following outline of the structure of mathematical abilities:

1. Obtaining mathematical information
    A. The ability for formalized perception of
        mathematical material, for grasping the
        formal structure of a problem.
2. Processing mathematical information
    A. The ability for logical thought in the sphere of
        quantitative and spatial relationships, number and
        letter symbols: the ability to think in mathematical
        symbols.
    B. The ability for rapid and broad generalization of
        mathematical objects, relations, and operations.
    C. The ability to curtail the process of mathematical
        reasoning and the system of corresponding opera-
        tions; the ability to think in curtailed structures.
    D. Flexibility of mental processes in mathematical
        activity.
    E. Striving for clarity, simplicity, economy, and
        rationality of solutions.
    F. The ability for rapid and free reconstruction of
        the direction of a mental process, switching from a
        direct to a reverse train of thought (reversibility
        of the mental process in mathematical reasoning).
3. Retaining mathematical information
    A. Mathematical memory (generalized memory for
        mathematical relationships, type characteristics,
        schemes of arguments and proofs, methods of
        problem-solving, and principles of approach).
4. General synthetic component
    A. Mathematical cast of mind.

He also lists five qualities which are 'not obligatory for
mathematical giftedness':

1. The swiftness of mental processes;
2. Computational abilities (abilities for rapid and precise
    calculations, often in the head);
3. A memory for symbols, numbers and formulae;
4. An ability for spatial concepts;
5. An ability to visualize abstract mathematical relation-
    ships and dependencies.

To give meaning to this structure it is necessary to consider
the distinctive ways in which mathematically gifted children
solve particular problems. For example, 'the ability to
curtail the processes of mathematical reasoning' (2C) is
illustrated by the response of Gilya, an 11-year-old boy, to
the problem: 'A father is 35 years old, and his son is 2. In
how many years will the father be four times as old as his
son?'. The problem can be solved algebraically (e.g.,
$4(2 + y) = 35 + y$, etc.) or by trial and error (perhaps with
the aid of a diagram) but Gilya simply reasons thus: 'He is
33 years older, that means one part will be 11, and the son
still needs 9 more years. In 9 years'. Gilya also solves
problems at a highly generalized, formal level (1A, 2A, 2B).

228

Ironically this prevents him noticing that the following problem is unrealistic: 'Represent the general form for numbers that leave a remainder of 7 when divided by 5'; nonetheless, his method of solution is impressive: 'In general, in these (sic) cases we must take the multiplier x for the given number and add the remainder so that it is divisible by y and there will be z in the remainder ... All the numbers will divide by $2y + z$. In the given case it will be $x.5 + 7$'.

In his experiments Krutetskii devised 26 series of problems, from which he constructed 79 tests (22 in arithmetic, 17 in algebra, 25 in geometry and 15 others). The first nine series are summarized below.

Series I - Problems with an Unstated Question: These were designed to test whether the child could 'perceive the logic of the relations and dependencies given in the problem, whether he understood their essence'.

> e.g., Twenty-five pipes of lengths 5 m and 8 m were laid over a distance of 155 m. (How many pipes of each kind were laid?)
> e.g., A man has lived x months. (How many years old is he?)

Series II - Problems with Incomplete Information: This series, like Series I, tests whether the child can perceive the structure of a problem.

> e.g., A train consists of tank cars, freight cars and flatcars. There are 4 fewer tank cars than flatcars, and 8 fewer tank cars than freight cars. How many tank cars, freight cars and flatcars does the train have? (Their total number is unknown.)
> e.g., Given two circles. The radius of one is 3 cm, and the distances between their centres is 10 cm. Do the circles intersect? (We must know the radius of the other circle.)

Series III - Problems with Surplus Information: Again the series is directed to the perception of problem structure.

Series IV - Problems with Interpenetrating Elements: These were all geometric problems concerned with notions akin to Witkin's 'field-dependence/field-independence'.

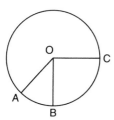

How many sectors are shown (and what are they)?

Series V - Systems of Problems of a Single Type: Here the child was presented with problems of the same form but increasing in complexity, to determine how quickly the form could be generalized. The items below would be presented in the order 8, 1, 8, 2, 8, 3, etc., with, in addition, interjections for the second list (10, 2, etc.) so as to see whether the child could differentiate between the two forms, despite the superficial similarity of the items.

  A. Algebra test

1. $(a + b)^2 =$
2. $(1 + \frac{1}{2}a^3b^2)^2 =$
3. $(-5x + 0.6xy^2)^2 =$
4. $(3x - 6y)^2 =$
5. $(m + x + b)^2 =$
6. $(4x + y^3 - a)^2 =$
7. $51^2 =$
8. $(C + D + E)(E + C + D) =$

1a. $a^2 + b =$
2a. $(\frac{1}{3} ab^3)^2 + (2a)^2 =$
3a. $(-5x^2 - 0.6xy^2) \cdot 2 =$
4a. $(3x + 6y) \cdot 2x =$
5a. $2(m^2 + x^2 + b^2) =$
6a. $4x^2 + y^2 - a^2 =$
7a. $98^3 =$
8a. $(c + m + x)(c - m + x) =$

Series VI - Systems of Problems of Different Types:

Series VII - Systems of Problems with Gradual Transformation from Concrete to Abstract:

  e.g., 1a The length of a room is 6m, its width is 3m and its height is 3m. What is the volume of 4 such rooms?

    1e The length of a room is β m, its width and height are α m each. What is the volume of η such rooms?

    2e A slab was cut into m squares of side x cm and n squares of side y cm. There were no scraps left. What is the area of the slab?

Series VIII - Composition of Problems of a Given Type: This series was to test whether the child could generalize from a *single* instance. Given one problem the child was asked to make up problems of the same type and of another type.

  e.g. Two workers, working 9 hours, made 243 parts. One of them made 13 parts in 1 hour. How many did the other make in 1 hour?

Series IX - Problems on Proof. These consisted of a number of tests, each involving increasingly complicated proofs of a simple type. The aim was to see whether, having understood the simplest problem, the child could transfer the underlying principle to more complex problems. For the algebra items below, the order of presentation would be 1, 7, 2, 7, 3, 7, 4 .... etc. Valuable and substantial reviews of Krutetskii's book and of an earlier set of Soviet studies have been written by van Bruggen and Freudenthal (1977a, b).

A. Algebra test

1. Can it be said that —d always expresses a negative number? Or that $2a > a$? Or that $8n$ must be an integer?
2. Under what conditions does: $a = 5a$; $a = a^2$; $a < a^2$; $a > a^2$; $-\dot{a} > 0$; $a - 2 > 0$; $a + b < a - b$?
3. For what value of m is the equality $am/bm = a/b$ invalid?
4. For what values of a and b does the equality $a/b = 3a/b$ hold? (The examinee should explore various possiblities.)
5. Prove that $x^4 + x + 1$ cannot be a negative number for any value of x.
6. Prove that the polynomial $x^{12} - x^9 + x^4 - x + 1$ is positive for all x.
7. Prove that for any a the expression $(a - 1)(a - 3)(a - 4)(a - 6) + 10$ will be a positive number.

One aim of an American project, *The Teaching of Ninth-Grade General Mathematics* (co-ordinated by Perry Lanier at the Institute for Research on Teaching at Michigan State University) is to investigate the extent to which some of the abilities delineated by Krutetskii are present or absent in 14-year-old school children who are not good at mathematics. To this end, questions have been devised involving curtailment, reversibility, flexibility and information gathering, examples of which are shown below.

Generalization: How many pear trees are there in an orchard which has 18 rows and 6 trees in a row? (The student is asked to explain *how* to solve this and two quite different problems, and then asked to invent a similar problem to this first one.)

Reversibility: Suppose you know $1376 + 30,805 = 32,181$. Find $32,181 - 1376$.
                    Solve $309 - 84$. Using that information, what other problems can you solve?

Flexibility: Add $27 + 3 + 13 + 7 + 21 + 9 + 1 + 19 =$ (The student is asked to solve the problem in as many different ways as he can find, and to state which method he likes the best.)

Information Gathering: (incomplete information)   A basketball player made 12 baskets in a game. What was her percent of successful shots to unsuccessful ones?

Thus sets of questions based in simpler arithmetic as compared with Krutetskii's rather more sophisticated, mainly algebraic material, have been prepared and a considerable number of pupils interviewed with them. These have produced very interesting and valuable insights, mainly showing the lack of the Krutetskii abilities in these pupils, but the

231

intention is to use these as the basis of some teaching in this grade. Thus the teaching methods emerging are those being adopted in the Nottingham SSRC teaching methods project, that is, the notions of drawing diagrams, reversing problems, moving from materials to symbolizations, generalization, the replacement of hard numbers by easy ones, as principles for promoting mathematical learning which the pupil may be able eventually to use consciously for himself.

7. MATHEMATICAL INVESTIGATION IN THE CLASSROOM AND ITS EVALUATION

Substantial progress has been made in Britain in developing ways of teaching mathematics which give the pupil a more genuine experience of mathematical activity than is obtained from a regime designed simply with a view to efficiency in learning particular concepts and skills. The pages of the journal *Mathematics Teaching,* in particular, contain many reports of such activities and suggestions for starting points for new investigations. For readers less well acquainted with this work, we list some of the more substantial materials. The ATM book *Some Lessons in Mathematics* (Fletcher, 1964) took pupil involvement in investigation for granted. The ATCDE booklet *Teaching Mathematics* (1967) reported work developed at some colleges of education: it had chapters entitled The Student's Mathematics and The Teacher's Mathematics. More recently, an ATM booklet *Maths for Sixth Formers* (ATM, 1979) outlined a possible course for this level which emphasised activities of Generalisation and Modelling, alongside the learning of standard concepts and skills. The Leapfrogs group has published a series of Action Books and other material intended to introduce 9-13-year-olds to explorations of a mathematical kind in areas of experience designed to appeal, e.g., *Codes, Networks, Spirals,* Pegboard games. *The South Nottinghamshire Project* (Bell, Rooke and Wigley, 1978, 1979) has produced material for the first two years of the secondary school which combines an emphasis on developing general mathematical strategies with the treatement of the normal subject matter. The strategies identified here are:

Generalization:

1. generating examples - to satisfy given conditions or to test a given conjecture;
2. classify and order systematically to obtain a complete set;
3. recognize, and (4) extend a pattern or relationships, numerical or spatial.
5. express a relationship in general terms, algebraically or verbally ('Make a generalization').

Proof:

1. Check all cases.
2. Establish sub-classes and check exhaustively.
3. Identify underlying general relationship ('Key fact').
4. Connect data and conclusion.
5. Embed in existing knowledge.

Formulating problems

Representation

1. Use of diagrammatic recording, graphs, tables.
2. Use of algebraic symbolism.

Abstraction:

1. Actions with concrete embodiments.
2. Abstractions resulting from generalization.

In connection with this project, tests designed to measure pupils' progress in acquiring these strategies have been used in the schools concerned. Comparisons with other schools using this test and a short number test gave reasonable grounds for concluding that strategies *were* being acquired without significant loss of content learning. (Bell, 1976, Ch.5). These tests have been subjected to further development and now contain sub-scales for Generalization, Classification, Exploration, Symbolization and Modelling (Horton, 1979).

In most schools, attempts to broaden the curriculum in this way draw to a close as the 16+ examinations approach; but there is at least one school (Wyndham, Cumbria) where such work is assessed for CSE and O-level, under special syllabus arrangements. For fuller description of current work of this type see Bell, 1979.

8. CONCLUSIONS AND IMPLICATIONS

A comparison of the results of the various studies reviewed above leads to the following tentative conclusions. First, the successful attempts to teach *general* problem-solving strategies were in the main those which involved intensive long-term explicit teaching of a few very general strategies. This was generally done in one context or one type of situation but the skills did seem transferable more widely. The Covington and Crutchfield programme and the Suchman Inquiry Training were of this kind. We may note also that these were both experiments with relatively young pupils, 10 to 11 years of age. (We do not know, however, what the general level of ability was.) The experiment of Lucas (1974) and of Schoenfeld (1978) suggest that periods of training as short as one term may also have an effect for students of college age and ability. The second conclusion is that drawing attention to the general characteristics in *particular areas* could produce more limited but effective transfer; for example, in the Kantowskii studies and those of Wills and Brien, general methods for the solution of problems involving generalization over values of *n* were learnt, and in the Kantowskii work the use of tables for logic problems was also picked up and used. (The Wills study was a relatively intensive training and the Kantowskii work was with able students.) Thirdly, we may note that the unsuccessful experiments were those which attempted to teach pupils of high school age (12 and 15) the general Polya scheme of heuristics, using a teaching package, by a relatively large and unselected group of teachers.

We suggest that the implication of this work is that, at all stages in mathematical work, awareness of problem types, reflection on and categorization of methods of solution, and this is being done in progressively wider areas of mathematics, thus considering questions of greater gener-ality, should be a feature of the learning. The present efforts in various quarters to increase the amount of invest-igation work in the mathematics classroom, efforts to in-crease the amount of real problem-solving being performed, and attempts to train students in the ability to construct and appreciate proofs need all to be seen as part of this development of general concepts and strategies.

References

ATCDE (1967): *Teaching Mathematics: Main Courses in Colleges of Education*. London: Association of Teachers of Colleges and Departments of Education.

ATM (1979): *Maths for Sixth Formers*. Nelson: Association of Teachers of Mathematics.

BELL, A.W. (1972-4): 'Proof in transformation geometry', (5 articles), *Maths Teaching*.

BELL, A.W. (1976): 'A study of pupils' proof-explanations in mathematical situations. *Ed Stud Math*, 7, 23-40.

BELL, A.W. (1978): 'Proof'. In: DÖRFLER and FISCHER (Eds): *Beweisen im Mathematikunterricht*. Stuttgart: Teuber.

BELL, A.W. (1979): 'The learning of process aspects of mathematics', *Ed Stud Math*, 10, 361-87.

BELL, A.W., ROOKE, D. and WIGLEY, A. (1978-9): *Journey into Maths: The South Nottinghamshire Project. Teacher's Guide and Pupils' Materials, Stages 1 and 2*. Blackie.

BELL, M. and USISKIN, Z. (1980): *A Conceptualisation of the Applications of Arithmetic*. Research report to NCTM. Department of Education, University of Chicago.

BRIAN, R.B. (1966): 'Processes of mathematics: a definitional development and an experimental investigation of their relationship to mathematical problem-solving behaviour', *Diss Abs.Int*, 67-11, 815.

BURTON, L. (1980): 'The teaching of mathematics to young children using a problem solving approach', *Ed Stud Math*, 11, 43-58.

CAMBRIDGE (MASS) CONFERENCE ON SCHOOL MATHEMATICS (1963): *Goals for School Mathematics*. New York: Houghton Mifflin.

COVINGTON, M.V. and CRUTCHFIELD, R.S. (1965): 'Facilitation of creative problem solving', *Programmed Instruction*, 4.

FREUDENTHAL, H. (1973): *Mathematics as an Educational Task*. Dordrecht: Reidel.

HORTON, B. (1979): *Tests of Mathematical Process*. Shell Centre for Mathematical Education, University of Nottingham.

KANTOWSKI, M.G. (1979): *The Use of Heuristics in Problem Solving: An Exploratory Study*. Technical report, National Science Foundation Project, University of Florida.

KILPATRICK, J. (1969): 'Problem solving in mathematics', *Rev Ed Res*.

KING, I.L. (1973): 'A formative development of an elementary school unit on proof', *J Res Math Ed*, 4, 1.

KRUTETSKII, V.A. (1976): *The Psychology of Mathematical Abilities in Schoolchildren*. University of Chicago Press.

KÜCHEMANN, D.E. (1977): *Secondary School Children's Understanding of Reflection and Rotation*. In: HART, K. (1981): *Children's Understanding of Mathematics, 11-16*. London: Murray.

LAWSON, A.E. and WOLLMANN, W.T. (1975): *Encouraging the Transition from Concrete to Formal Cognitive Functioning - an Experiment*. AESOP, Lawrence Hall of Science, Berkeley, California.

LOMON, E. *et al.* (1975): 'Real problem solving in USMES: interdisciplinary education and much more', *Sch Sci Math*.

LUCAS, J.F. (1974): 'The teaching of heuristic problem-solving strategies in elementary calculus', *J Res Math Ed*, 5, 1.

MATHEMATICAL ASSOCIATION (1923): *The Teaching of Geometry in Schools*. London: Bell.

MAXWELL, E.A. (1975): *Geometry by Transformations*. Cambridge University Press.

MUELLER, D.J. (1975): Logic and the ability to prove theorems in geometry. Ph.D. thesis, Florida State University. *Diss Abs Int,* 36A, August, 851.

NEWELL, A. and SIMON, H.A. (1972): *Human Problem Solving*. Prentice Hall, Englewood Cliffs.

PHILLIPS, H.P. (1968): A comparative study of the effectiveness of two methods of teaching elementary mathematical proofs. Ph.D. thesis, Ohio State University.

POLYA, G. (1945): *How to Solve It*. Princeton University Press.

POST, T.R. (1967): 'The effects of the presentation of a structure of the problem-solving process upon problem-solving ability in seventh grade mathematics', Ph.D. thesis, Indiana University. *Diss Abs Int,* 28, 4545A.

QUADLING, D. (1972): *SMP, The First Ten Years*. Cambridge University Press.

REYNOLDS, J. (1967): The development of the concept of proof in grammar school pupils. Ph.D. thesis, University of Nottingham.

ROBINSON, G.E. (1964): An investigation of junior high school students' spontaneous use of proof to justify mathematical generalizations. Ph.D. thesis, University of Wisconsin, Madison.

ROY, G.R. (1970): The effect of the study of mathematical logic on student performance in proving theorems by mathematical induction. Ph.D. thesis, State University of New York, Buffalo.

SCHOENFELD, H.H. (1979): 'Explicit heuristic training as a variable in problem solving performance', *J Res Math Ed,* 10, 173-87.

SMP: (1969-74):*SMP Books A-H, X, Y, Z*. Cambridge University Press.

SCOTTISH MATHEMATICS GROUP (1965-68 1st ed.), (1971-74 2nd ed.): *Modern Mathematics for Schools, Books 1-7*. London: Blackie and Chambers.

SCOTT, N. (1973): 'Cognitive style and inquiry strategy: a five year study', *J Res Sci Teaching,* 10, 323-30.

SCOTT, N. (1977): 'Enquiry strategy and mathematics achievement', *J Res Math Ed,* 8, 132-43.

SHANN, M. (1974): *Measuring Problem Solving.* School of
Education, Boston University.

SHAPIRO, B. (1975): *see* LOMON, E., *et al.* 'Real problem
solving in USMES: interdisciplinary education and much more',
Sch Sci Math.

SUCHMAN, J.R. (1961): 'Inquiry training: building skills for
autonomous discovery', *Merrill-Palmer Quarterly,*
147-69.

TREILIBS, V., LOW, B. and BURKHARDT, H. (1979): *Formulation
Processes of Mathematical Modelling.* Shell Centre for
Mathematical Education, University of Nottingham.

USISKIN, Z. (1980): *A Test of the Relevance of the Van Hiele
Level Theory with Respect to the High School Geometry Course.*
Research report to NCTM. Department of Education, University
of Chicago.

VAN BRUGGEN, J.C. and FREUDENTHAL, H. (1977a): *A Review of
Soviet Studies on the Psychology of Learning and Teaching
Mathematics* (6 volumes). Stanford, California: National
Academy of Education.

VAN BRUGGEN, J.C. and FREUDENTHAL, H. (1977b): *The Psychology
of Mathematical Abilities in Schoolchildren.* Stanford,
California: National Academy of Education.

VAN DORMOLEN, J. (1977): 'Learning to understand what giving
a proof really means', *Ed Stud Math,* 8, 27-34.

VAN HIELE, P.M. (1959): 'La pensée de l'enfant et la
geometrie', *Bulletin de l'Association des Professeurs
Mathematiques de l'Enseignement Public,* 198.

VAN HIELE, P.M. and VAN HIELE-GELDOF, D. (1958): 'A method
of initiation into geometry', In: FREUDENTHAL, H. (Ed):
Reports on Methods of Initiation into Geometry. Groningen:
Wolters.

WILLIAMS   (1978): An investigation of senior high school
students' understanding of the nature of mathematical proof.
Ph.D. thesis, University of Alberta, Edmonton.

WILLS, H. (1967): 'Transfer of problem solving ability gained
through learning by discovery', Ph.D. thesis, University of
Illinois. *Diss Abs Int,* 67-11, 937.

WILSON, J.W. (1967): Generality of heuristics as an
instructional variable. Ph.D. thesis, Stanford University.

WIRSZUP, I. (1976): 'Breakthroughs in the psychology of
learning and teaching geometry', In: MARTIN, J.L. and
BRADBARD, D. (Eds): *Space and Geometry*. Columbus, Ohio: ERIC.

WYNNE WILLSON, W. (1977): *Geometry*. (For the Schools Council:
The Mathematics Curriculum, A Critical Review.) London:
Blackie.

Chapter Nine
Attitudes

1. Introduction

2. Attitudes to Mathematics as a Whole

3. Attitudes towards Different Topics within Mathematics

4. Perception of the Nature of Mathematics

5. The Relationships between Attitudes and Attainment

6. The Effect of Attitude on the Choice of Mathematics for Further Study

7. Summary of Conclusions

1. INTRODUCTION

Most teachers quite rightly attach considerable importance to the promotion of favourable attitudes in their mathematics classes. It is generally felt that liking for, and interest in mathematics lead to greater effort and hence to higher achievement, and also to the willingness to pursue mathematics in subsequent studies. It is also the case that one of the motivations for the development of modern mathematics courses in the 1960s was the wish to make school mathematics more interesting and enjoyable and more representative of mathematics as it is explored and used in the world outside school. In this chapter we shall review the considerable amount of research which exists on pupils' attitudes to mathematics, both in Britain and the United States, and consider what light it throws on these questions. It is possible to identify certain main areas of investigation from this literature: this, however, does not provide a classification of the research - most studies include aspects of at least two of these areas.

1. Attempts to investigate and describe children's *attitudes to mathematics as a whole*: There are several dimensions to this: like - dislike, easy - hard, useful - not useful are the most common. Some studies of this type compare attitudes to a number of subjects, so that mathematics

appears in a kind of league table, according to one or more criteria.

2. Comparison of *attitudes towards different topics within mathematics:* The dimensions and rationale here are similar to (1), but are concerned with different topics within the subject, rather than comparisons amongst subjects.

3. Investigations of pupils' *perception of what mathematics is:* Some research, rather than recording fairly immediate affective reactions to mathematics, has tried to probe more deeply into notions of what sort of notion pupils have of the nature of mathematics. Thus, the subject might be considered as a body of received knowledge, or as a (possibly creative) activity, or as a set of tools for tackling practical problems.

4. Discussions of the *relationship between attitude and achievement:* There is an interaction between attitude and achievement, and some studies are concerned with measuring the degree of correlation between these pupil attributes. Some work, however, apparently attempts to discover a causal relationship. On the one hand, successful performance might be listed as a factor in developing favourable attitudes: on the other, certain identifiable attitudes may enhance or inhibit achievement.

5. Attempts to discover *how attitudes influence pupils in choosing mathematics for further study:* What is it that determines whether a student will opt to take mathematics in the sixth form or at university?

2. ATTITUDES TO MATHEMATICS AS A WHOLE

Sharples (1969) has compared children's attitudes to various subjects in junior schools, by considering their reactions to the following statements of attitude towards mathematics, reading, writing stories, art and physical education in turn:

    (1)   I hate it;
    (2)   It is the worst thing we do in school;
    (3)   I can't stand it;
    (4)   It is all right sometimes;
    (5)   I think it is good;
    (6)   It is most enjoyable;
    (7)   It is good fun and I like it very much;
    (8)   I love it.

Pupils indicated agreement with statements by a tick: the score recorded is either the lowest numbered item ticked below (4) or the highest above (4) - it is not possible to accept items on both sides of (4).

This was carried out in four schools each of which had a

240

curriculum 'bias' in favour of a particular activity (P.E., art, maths and language respectively). In all cases, mathematics occupied a low position (with respect to attitude) compared with other subjects: it was more favoured in the school which placed most emphasis on it, but even in this school it was in the next to the last position - only 'writing' was disliked more. In the other three schools, mathematics was clearly last. All four schools were free from 11+ selection, so there was no pre-11+ pressure involved.

Sharples' results can also be analysed by sex and age (9, 10 and 11), but no significant difference emerges amongst subject preferences. However, attitudes towards all five activities become less favourable from the age of nine to the age of 11. This does not appear to be related to the teaching approach: attitudes amongst the older children were less favourable to all activities, some of which became less 'child-centred' and some of which (including art and physical education) patently did not.

A comparable analysis was carried out by Greenblatt (1962) in California; children were asked to list in order of preference their three favourite school subjects. The conclusions are very different from Sharples', but it is doubtful whether asking children to produce a list in this way is an appropriate approach. Thus, Inskeep and Rowland (1965), questioning the ability of pupils to list preferences, have shown that not only the fact of presenting a list for children to choose from, but also the order of presentation has a significant effect.

Amongst other American studies, the work of Levine (1972) is worth mentioning. The study obtains rankings of four subjects according to the responses of pupils and their parents to nine statements, which reflect their perceptions of the importance of mathematics, their own interest and ability in the subject, and teachers' competence in mathematics. Over all, the results and discussion suggest that pupils look upon mathematics in a favourable light compared with the other three subjects quoted (English, Science, Social Studies); although, of course, 'favourable' here involves a variety of attitudes rather than the specific 'like-dislike' dimension of Sharples' work.

A thesis by Gopal Rao (1973) extends the investigation of British attitudes to mathematics to secondary school pupils, and also considers some of the factors which might create and affect attitudes. Mathematics is often liked and often strongly disliked in primary schools, there is considerable polarization and fairly definite attitudes are established by the age of 11. Attitudes seem to become generally less favourable during the early years of secondary school, stabilizing again by about age 14; but this may merely

241

reflect a continuation of the general decline in attitudes towards all school subjects, noted by Sharples. Although significant differences in attitude occur between particular schools these are not apparently related to the *type* of secondary school (grammar, comprehensive, etc.), nor to the syllabus content. There is significant correlation between attitude to school and to mathematics, significant but low correlation between parents' and children's attitudes to mathematics, and significant correlation with the peer-group's attitude. After school, boys generally have a more favourable attitude than girls - but all apparently believe mathematics to be useful.

Some comparable work in the United States is provided by Callahan (1971), who also observes a proportion of pupils expressing 'extreme dislike' and some expressing a strong liking for mathematics, in the 11 to 14 age range. He notes that lasting attitudes to mathematics can be developed at any age, but that the most important age for establishing these is about 11. He also discovers the general belief that mathematics is useful - 66 per cent of pupils feel that it is as important as (or more important than) any other subject - and its practical importance is frequently given by pupils as their reason for liking mathematics. It can be seen that Callahan's conclusions are in fairly close agreement with Gopal Rao's.

An attempt at a detailed comparison of secondary school pupils' attitudes to different subjects was made by Duckworth and Entwistle (1974). They used a 'repertory grid technique' in which pupils were presented with 20 pairs of contrasting comments and asked in each case to agree with one comment (A) or the other (B): a response (O) indicating uncertainty or neutrality was available, but its use was discouraged. The sample included second-form and fifth-form pupils in Lancashire, and the results are presented in the form of 'league tables' of nine subjects according to 'interest, difficulty, freedom and social benefit' for both second-formers and fifth-formers. In 'interest', mathematics appears consistently low (around 6th/7th for both boys and girls). In 'social benefit' mathematics falls from second place (in second year) to fourth (in fifth year), again for both boys and girls. The precise details of this study can only be identified through consideration of the individual comments appearing in the grid overleaf.

3. ATTITUDES TOWARDS DIFFERENT TOPICS WITHIN MATHEMATICS

The work of Kyles and Sumner (1977) is concerned with general attitudes to mathematics, but also investigates the details of pupils' reactions to different topics and activities within the subject. They confirm that mathematics is considered useful by both primary and secondary pupils, the latter recognizing its importance for future employment; but the subject is not particularly liked in primary schools or

242

Repertory grid used to measure attitudes to school subjects

Comment A	English	Physics	History	Biology	French	Chemistry	Latin	Mathematics	Geography	Comment B	Scale	
1. Rather dull and monotonous										Can be exciting	I	B
2. Helps to satisfy my curiosity about life										Does not really satisfy my curiosity	I	A
3. My own ideas can be used										Not much room for my own ideas	F	A
4. Most pupils can do it quite well										Few seem able to do it well	D	B
5. Usually precise and exact										Often vague and woolly	not used	
6. Doesn't require too much work										Needs really hard work	D	B
7. Rather narrow and specialised										Of fairly general interest	I	B
8. A lot of learning by heart										Learning by heart not really important	F	B
9. Needs quite a lot of imagination										Imagination seldom required	F	A
10. Tends to be difficult										Fairly easy	D	A
11. Knowledge useful in everyday life										Knowledge not much use in everyday life	SB	A
12. Requires wide reading outside lessons										School-books and lessons enough	F	A
13. Usually interests me										Often bores me	I	A
14. We don't spend enough time on it										Enough time on the whole	not used	
15. Important for solving world problems										Not particularly important for solving world problems	SB	A
16. Gives opportunity to think things out for myself										Too much of other people's knowledge	F	A
17. Tends to be complicated										Generally straightforward	D	A
18. I enjoy it more than I used to										My liking for it has decreased	I	A
19. Knowledge of it helps people to understand one another										Knowledge of it doesn't seem to help in understanding people	SB	A
20. Facts and ideas hard to grasp										Facts and ideas not really difficult	D	A

Key to scales: I—Interest. D—Difficulty. F—Freedom. S—Social Benefit.

Method of scoring
The second letter in the final column above indicates the response which is scored though, in practice, the 'Difficulty' dimension is better scored as 'Ease.' The 'scoring' poles are thus reversed. Agreement with the indicated pole is scored as '2,' 'O' responses are scored as '1,' disagreement as zero. The maximum score on each of the three main dimensions is, therefore, 10 and the minimum is zero.

found absorbing by secondary pupils. 'Anxiety' is a relevant factor in the primary age group, about half the pupils being affected by this to some degree.

In their analysis of affective response to particular topics and tasks, Kyles and Sumner consider two dimensions, 'easiness' and 'liking'. In 'easiness', children's views agreed closely with those of their teachers. In order of decreasing easiness, operations appear in the order -, +, x, ÷ and topics in the order: whole numbers, decimals, fractions. These results were derived by presenting children with workcards and asking them specifically to choose which workcard would be easier for them to do. By this means, each card was given an 'easiness score': examples and their relative positions on the scale are shown in the diagram overleaf.

In terms of 'liking', word problems are generally more popular than other examples; a workcard involving a word problem concerned with money consistently topped the popularity scale. In general, it also appears that children would much rather do number geometry questions than algebra ones: primary school children respond to algebra questions with comments such as 'I don't like finding numbers' and 'I don't like missing numbers'. 'Multiplying decimals' also appears consistently higher on the liking scale than 'subtracting decimals'.

Pupils' attitudes to specific topics are also identified in the APU Primary Survey (1980). In general this survey again reveals a strong tendency for pupils to find mathematics useful, but children appear more qualified in other aspects of attitude toward mathematics. When asked whether they enjoyed mathematics, the answer was 'sometimes': liking and difficulty are not easily attributable to the whole subject - they are associated with specific topics and forms of presentation. This makes the discussion of particular topics relevant and important.

A number of features seem worth noting:

1. 'Modern' geometry topics (symmetry, tessellations) appear less liked than those in descriptive geometry (shapes, angles);

2. In 'measures', money, weighing and time are generally more popular than length, area and volume;

3. The 'number' category is split into three clear groupings by results. Number computations are the most popular, receiving more liking responses than any other topic category. Next come the questions concerned with fractions, decimals and percentages; and least popular

Examples of Comparisons of Workcards for 'Liking'

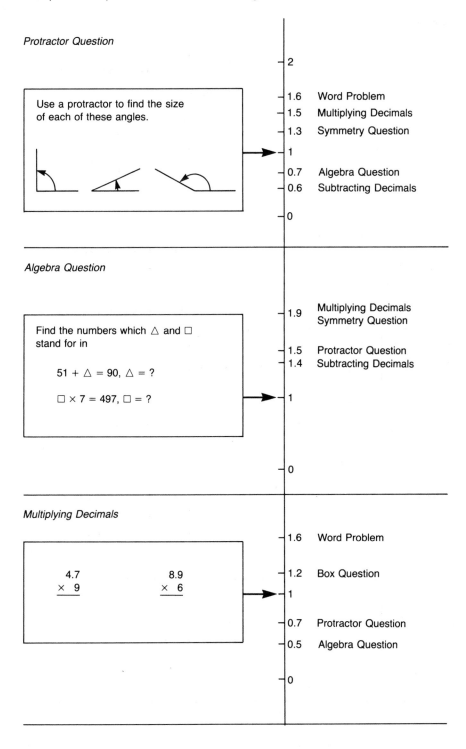

Protractor Question

Use a protractor to find the size
of each of these angles.

2

1.6 Word Problem
1.5 Multiplying Decimals
1.3 Symmetry Question
1
0.7 Algebra Question
0.6 Subtracting Decimals
0

Algebra Question

Find the numbers which △ and ☐
stand for in

$51 + △ = 90, △ = ?$

$☐ × 7 = 497, ☐ = ?$

1.9 Multiplying Decimals
     Symmetry Question
1.5 Protractor Question
1.4 Subtracting Decimals
1
0

Multiplying Decimals

4.7   8.9
× 9   × 6

1.6 Word Problem
1.2 Box Question
1
0.7 Protractor Question
0.5 Algebra Question
0

Examples of Workcard Comparisons for 'Easiness'

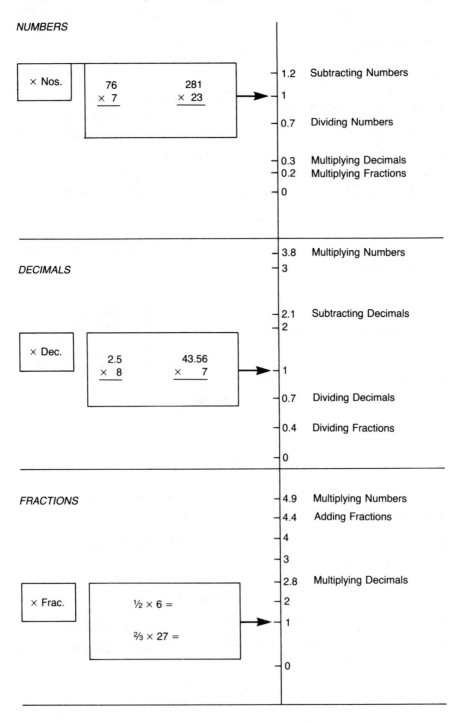

NUMBERS

× Nos.

$$\frac{76}{\times\ 7}\qquad\frac{281}{\times\ 23}$$

1.2   Subtracting Numbers
1
0.7   Dividing Numbers

0.3   Multiplying Decimals
0.2   Multiplying Fractions
0

DECIMALS

3.8   Multiplying Numbers
3

× Dec.

$$\frac{2.5}{\times\ 8}\qquad\frac{43.56}{\times\ 7}$$

2.1   Subtracting Decimals
2

1

0.7   Dividing Decimals

0.4   Dividing Fractions

0

FRACTIONS

4.9   Multiplying Numbers
4.4   Adding Fractions
4

3

2.8   Multiplying Decimals

× Frac.

½ × 6 =

⅔ × 27 =

2

1

0

246

are items involving number concepts (factors, prime numbers).

Although the APU survey shows little general difference in attitudes between boys and girls, they do differ in response to particular topics. Girls are generally more apt to tick 'unsure' than boys: this is a consistent pattern throughout topics, but shows up most strikingly in the 'measures' category. In most of the responses, sex differences appear to point to girls' uncertainty towards topics rather than an active dislike or difficulty factor.

4. CHILDREN'S PERCEPTION OF THE NATURE OF MATHEMATICS

In a research study, Preston has considered pupils' affective behaviour in following secondary mathematics courses. He identifies (through factor analysis) three definite, named, facets of behaviour:

A. tendency to see mathematics as an algorithmic, mechanical and somewhat stereotyped subject;

B. tendency to see mathematics in an open-ended, intuitive and heuristic setting;

C. commitment, interest and application to mathematics.

The results are analysed in various ways.

1. The mean score of boys is significantly higher than that of girls for factors B and C: Preston states that 'girls see mathematics set in a rather restricted and predictable environment. Their level of interest and commitment is lower than for boys.'

2. City children have a higher mean score for factor A than do suburban or rural children, suggesting that the former see mathematics in a more mechanical and stereotyped way. Scores for factor C are higher for suburban/semi-rural children than for city children, but the responses from a totally rural setting show the lowest level of this factor.

3. The results of pupils taking a modern (SMP) course differ significantly from the norm. The high level of factor B mean score indicates that these pupils see mathematics in a wider context of application, that they have a more strongly developed sense of intuition and their approach to problems allows greater flexibility. Disappointingly, however, the factor C mean score for this group, showing level of commitment and interest, is significantly below the norm.

247

Breakdown of pupils' reponses to each topic in reference to Liking and Difficulty.
L = Like, D = Dislike, E = Easy, H = Hard, U = Uncertain, N = Not done. (Figures
represent the percentages of all pupils who took the Attitude Questionnaire.)

Percentages

Main content category of curriculum framework	Topic terms	Response category							
		Liking				Difficulty			
		L	D	U	N	E	H	U	N
Geometry	Geometry	25	12	18	41	20	15	21	41
	Shapes	71	11	14	3	73	6	15	4
	Angles	46	30	19	4	43	29	24	3
	Triangles	54	17	22	5	53	15	23	7
	Symmetry	35	13	17	32	34	12	19	33
	Tessellations	23	8	14	52	23	9	14	51
Measures	Measuring	68	16	11	2	68	12	16	2
	Money	81	7	8	1	85	5	7	2
	Weighing	55	19	19	6	52	15	27	4
	Time	70	13	13	2	76	9	11	2
	Length	60	18	17	2	63	14	19	3
	Area	50	27	18	4	47	26	21	4
	Volume	22	22	20	34	20	22	23	34
Number (concepts)	Number bases	23	18	21	37	20	19	22	37
	Factors	26	23	23	27	25	22	25	27
	Prime numbers	30	22	24	22	31	18	27	22
	Number patterns	47	12	21	18	44	11	23	19
Number (concepts/skills)	Fractions	54	29	14	2	47	29	20	2
	Decimals	47	29	18	5	44	30	20	5
	Percentages	37	24	18	19	35	26	19	18
	Multiplication tables	64	18	13	2	63	16	17	2
Number (computations)	Arithmetic	53	14	20	10	48	14	23	12
	Adding	92	3	3	1	94	2	3	0
	Subtracting	81	8	7	1	84	5	6	2
	Multiplying	76	12	9	1	73	12	11	2
	Dividing	69	18	10	1	69	14	14	1
Number (applications)	Problems	42	26	19	11	33	31	22	12
Algebra	Algebra	20	8	12	58	15	10	15	58
	Equations	20	14	20	45	15	18	20	44
	Sets	51	14	21	12	51	10	22	14
	Sorting	41	13	20	24	43	10	21	24
Probability and statistics	Graphs	74	9	12	4	69	8	18	4
	Averages	36	20	22	21	35	18	23	22
	Practical maths (not in a main content category)	54	12	22	10	48	11	27	12

Conclusion (3) is both unusual and important. A just-
ifiable criticism of curriculum change is that there is
little evidence that the new is superior to the old: research
and results such as are quoted here provide the kind of
information needed to evaluate innovation, particularly when
this represents a widespread shift in both content and
teaching approach. However, it needs to be remembered that
this research was done at a time when SMP was only used in a
relatively small number of schools, so that these represent
an untypical example.

In the APU Primary Survey, some of pupils' responses to
specific topics may suggest how they conceptualize mathematics. A possible way to investigate this is to consider
how pupils' over-all liking of mathematics is related to
their liking of particular topics and tasks. Pupils who
scored higher on the Liking of Mathematics scale liked all
four arithmetical operations, while the lowest scorers on
the Liking of Mathematics scale disliked the more difficult
operations (particularly division) much more than the easiest.
Morevover, in taped discussion, pupils frequently named
addition, subtraction, multiplication and division as the
most important topics in mathematics, while averages and
percentages were considered the least important.

The inclusion of a factor named 'freedom' in the
previously mentioned study by Duckworth and Entwistle gives
some indication of pupils' perception of mathematics.
Earlier work, described in this article, by which the grid
structure was determined, identified a factor clearly distinct
from 'interest' and 'difficulty', which involved use of
imagination and the pupil's own ideas. It was found that
scores for English were particularly high for this factor;
and a group of pupils expressed the view that this subject
provided 'freedom to express one's own ideas'. On this factor,
mathematics appears in seventh and fifth place for second-
year boys and girls respectively. By the fifth year of
secondary school, mathematics is still ranked fifth by girls
but has risen to fourth place for boys.

In an attempt to clarify the different ways in which
children perceive mathematics, Erlwanger (1974) has made
detailed case studies of a small group of children. He
considers that their expressed ideas, beliefs, emotions and
views about mathematics and learning mathematics are part
of their underlying conception of the nature of the subject;
and that this conception is a stable, cohesive system.
Further, this system is complex and varies considerably
between one child and another. Thus Erlwanger is concerned
with describing the general trend and direction of each
individual child's thinking, which underlies his observable
behaviour. Thus one child was apparently independent and
confident in mathematics, and it was found that this was
attributable to a belief that he was able to develop his

own system of rules: this made him both secure and independent. Another seemingly independent and confident child, however, had an underlying belief about mathematics which made him completely dependent on his booklets, and on 'sticking to the rules'.

A detailed example of the way Erlwanger carries out and analyses his case studies is provided by his description of one child's conception of rules and answers. A 12-year-old boy, Benny, was making good progress in a structured mathematics scheme (IPI - Individually Prescribed Instruction), and it was expected by the teacher that he could not have progressed so far without an adequate understanding and mastery of previous work. The study shows, however, that Benny's performance relies on underlying misconceptions in many areas, which are illustrated by his approach to various tasks including the following (Erlwanger, 1973):

ADDITION OF FRACTIONS:

Benny had already completed work on equivalent fractions, and addition of fractions with common denominators for $\frac{1}{2}$ through $\frac{1}{12}$. He appeared to understand halves and fourths, e.g., he knew that $\frac{1}{2} + \frac{1}{4} = \frac{3}{4}$. Benny believed that there were rules for different types of fractions, as illustrated by the following excerpt:

B: In fractions we have 100 different kinds of rules ....

E: Would you be able to say the 100 rules?

B: Ya ... maybe, but not all of them.

He was able to state addition rules for fractions clearly enough for me to judge that they depended upon the denominators of the fractions and were equivalent to the following:

$$\frac{a}{b} + \frac{c}{b} = \frac{a + c}{b}, \quad \text{e.g.,} \quad \frac{3}{10} + \frac{4}{10} = \frac{7}{10};$$

$$\frac{a}{b} + \frac{c}{d} = \frac{a + c}{b + d}, \quad \text{e.g.,} \quad \frac{4}{3} + \frac{3}{4} = 1;$$

$$\frac{a}{b} + \frac{c}{c} = 1\frac{a}{b}, \quad \text{e.g.,} \quad \frac{2}{3} + \frac{4}{4} = 1\frac{2}{3};$$

$$\frac{a}{10} + \frac{b}{100} = \frac{a + b}{110}, \quad \text{e.g.,} \quad \frac{6}{10} + \frac{20}{100} = \frac{26}{110}.$$

Benny had also used fraction discs ... when he showed me how he used them, he arrived at an incorrect result, as shown below:

E: Now when you simplify $\frac{3}{6}$ what do you get?

B: It should be $\frac{1}{2}$ because we got these fraction discs.

(But then he goes on to say) When you add $\frac{1}{4}$ and $\frac{1}{3}$

and $\frac{1}{8}$ equals $\frac{1}{2}$ (instead of $\frac{3}{15}$, as his rule for adding fractions, above, should give).

But fractions, to Benny, are mostly just symbols of the form $\frac{a}{b}$ added according to certain rules. This concept of fractions and rules leads to errors such as $\frac{2}{1} + \frac{1}{2} = \frac{3}{3} = 1$. Further, $\frac{2}{1} + \frac{1}{2}$ is 'just like saying $\frac{1}{2} + \frac{1}{2}$ because $\frac{2}{1}$, reverse that, $\frac{1}{2}$. So it will come out one whole no matter which way. 1 is 1."

The underlying perception is clearly that learning mathematics is merely applying rules to problems in order to get correct answers. Subsequent remedial work reported indicates that Benny's attitude and views can be changed: he is co-operative, responsive and eager to learn, but even so the process of re-forming developed conceptions is a very gradual one.

The essential differences in Erlwanger's case studies make general comments difficult - but this is part of their point. It is clearly suggested that children's distorted ideas and beliefs steadily develop and become more complex. Further, the evaluation and diagnostic procedures based on lists and brief conferences, which the teachers observed by Erlwanger employed, failed to detect children's underlying conceptions. This usually resulted in incorrect inferences about the nature of the children's understanding, success or failure.

5. ATTITUDES AND ACHIEVEMENT

There is surprisingly little research evidence to support the seemingly fairly reasonable belief that favourable attitudes towards mathematics lead to higher achievement in the subject. Jackson (1968), in a review of research into the way in which favourable attitudes to school, and school subjects generally, might enhance learning, concluded that nearly all investigations have found no significant relationships between attitudes and achievement. Knaupp (1973) has noted that in arithmetic we have little research basis for believing that these things are causally related. Aiken (1976) in his survey of work on attitudes towards mathematics does refer to some large-scale investigations which show significant correlation between attitude and achievement, but the correlations are still low. One such study is by Neale, Gill and Tismer (1970), who do indeed find significant correlation between attitude and achievement amongst lower-secondary-age pupils. However, the correlation is higher for boys than for girls, and the authors suggest that this may mean that achievement in arithmetic 'is somehow less rewarding to girls than to boys'. In contradiction to this interpretation, however, Greenblatt (1962) states that 'girls ... appear to choose or not choose arithmetic in terms of their achievement in the subject.'

It is somewhat difficult to be convinced that there is a significant difference between the sexes on this issue. Correlations of about 0.3 can be determined with precision only from a fairly large sample; and differences which indicate a correlation marginally above a certain significance level for one sex and just below for the other serve only to indicate a need for further evidence.

A summary of some studies relating attitude to achievement is given by Hart (1976), who comments that, even when a significant correlation occurs, it is difficult to determine whether the attitude to mathematics affects the achievement or vice versa. There may of course be other variables present which affect both attitude and achievement but which are not disclosed by statistical analysis. In her own research, although significant correlation between attitude and achievement is found for a sample of 179 pupils, Hart states that less than 20 per cent of the variance in attitude could be attributed to achievement variance: many children already had a negative attitude towards mathematics. Because the causal relationship is unclear, studies in this field need careful interpretation. Neale, Gill and Tismer find that differences in attitude can be significant predictors of achievement in arithmetic (not, it may be noted, in Science, Social Studies or Reading), but the amount of variation in achievement predictable by attitudes is very slight. Moreover, the word 'predict' here cannot be assumed to indicate that attitude affects achievement rather than vice versa. Neale, Gill and Tismer claim that their study is consistent with the suggestion that 'children learn in school not because they like to learn or value learning, but rather ... because the system compels them to learn, like it or not. This study also identifies the downward trend in pupils' attitudes towards mathematics particularly and towards school subjects generally.

All the evidence of age-related decline in self-reported attitudes should perhaps be reconsidered with both caution and open-mindedness. It is at least possible that children as they grow older tend to report less favourable (or more critical) reactions to a variety of experiences, and not merely to their school work. Such a systematic change in response could be a part of the maturing process, and might explain many observed 'downward' changes in attitude.

These changes are no doubt real but are not specific to the learning of mathematics.

6. THE EFFECT OF ATTITUDE ON THE CHOICE OF MATHEMATICS FOR FURTHER STUDY

The discussion so far has been concerned with attitudes as they are observed and established in primary and early secondary schools. Older pupils, however, themselves make a choice about whether to continue to study mathematics, first in the sixth-form and then in higher education. Their willingness to opt for mathematics is a well-defined characteristic, and two recent surveys, by Kempa and McGough (1977) and Stoodley (1979), have analysed this in the context of sixth-formers' self-reported attitudes towards the subject.

The object of Stoodley's study was to ascertain the pupils' attitudes towards seven possible types of mathematics degree courses (pure maths, applied maths, computer science, statistics, mathematical sciences, joint degrees and mathematics/education degrees), and also towards several features which degree courses might possess (single subject in depth, broad range, joint degrees with maths as a component, industrial placement year, general first year with speicalization later). Courses were to be rated on a five-point scale according to their (a) difficulty, (b) interest, (c) career prospects, (d) over-all attractiveness as measured by individual's willingness to take the course.

The over-all results show little correlation between either career prospect or difficulty and over-all attractiveness, but a strong correlation between interest and willingness to take the course. Boys were relatively more attracted by 'applied maths', girls by 'joint degrees' and 'mathematics/education degrees'.

A significant feature of the ratings is an apparent hardening of opinion against mathematics from lower to upper sixth. Pure mathematics in particular was rated very much lower on interest, more difficult and less attractive by the upper sixth; and, though to a lesser extent, so was applied mathematics.

Kempa and McGough (1977) investigated the attitudes of over 300 first-year sixth-form students in relation to the type of mathematics curriculum previously followed, to the choice of sixth-form subjects and to their mathematical achievement. The study involved pupils who had studied three different O-level JMB syllabuses, of which A and B are lumped together as 'traditional' and C is termed 'modern'. Attitudes were measured in terms of 'ease', 'usefulness' and 'enjoyment' and no difference is exhibited in these traits between the groups who had taken 'modern' and 'traditional' courses.

Thus, if one intention of 'modern' courses is to enhance enjoyment, this intention is apparently not being achieved. Of the attitude measures employed, liking/enjoyment of mathematics is the one which most strongly identifies students with a bias towards studying mathematics, suggesting that liking may be a stronger determinant of choice for further study than the relative absence of learning difficulties associated with the subject. Interestingly, students' self-reported opinions of mathematics are not closely dependent on their previous achievement in the subject: even their perception of its easiness or difficulty is only weakly related to attainment. Views about usefulness differentiate more between science/non-science specialists than between different biases towards further study in mathematics.

Another study of attitudes to mathematics among A-level students is provided by Selkirk (1972). In this study, the deterioration during the sixth-form years is again noted: there is evidence that some good calibre pupils are put off by too marked a change between their ordinary and advanced level courses. Again he finds no evidence to suggest that type of syllabus has any significant effect on attitude or performance. A particular point to which Selkirk draws attention is that very few pupils regard the teaching of mathematics as attractive to them: about two-thirds of those taking A-level mathematics were considering a degree course involving some mathematics, but only about 5 per cent were thinking of teaching the subject at a level above primary.

7. SUMMARY OF CONCLUSIONS

It is not easy to pick out points which summarize all the research on attitudes to mathematics. Strongly polarized attitudes can be established even amongst primary school children, and about 11 seems to be a critical age for this establishment. Attitudes are derived from teachers' attitudes (though this affects more intelligent pupils rather than the less able), and to an extent from parents' attitudes (though the correlation is fairly low). Attitude to mathematics is correlated with attitude to school as a whole (which is fairly consistent across subjects) and with the peer-group's attitude (a group-attitude tends to become established). These things do not seem to be related to type or size of school or to subject content.

Throughout school, a decline in attitudes to mathematics appears to go on, but this is part of a decline in attitudes to all school subjects and may be merely part of an increasingly critical approach to many aspects of life.

As far as attitudes to more specific topics within mathematics are concerned, we are able to identify certain topics as being generally liked and certain disliked; and some topics are generally considered 'easy', others 'hard'.

Underlying differences can be discovered in pupils' perceptions of the nature of mathematics: these may be determined by external and social factors. On the other hand, as in the Erlwanger studies, basic misconceptions may develop unchecked and subsequently lead to difficulties which are not easily remedied. This is a point on which action by a teacher is indicated. It suggests that there needs to be sufficiently open discussion in the classroom, not only about details of the subject-matter but also about the nature of mathematics in general, and of the nature and purpose of the work currently being undertaken, to ensure that gross misconceptions are exposed and if necessary. substantial shifts made in the general approach to the subject in the classroom in order to remedy these.

In particular, the research seems to show that modern courses do not, of themselves, lead to improved attitudes to the subject. It would seem important to look more closely at the nature of our classroom activities, trying to see them from the pupil's point of view, in order to understand their effect on the pupil's attitude and appreciation of the subject.

There appears to be an identifiable (although small) correlation between attitude and achievement: it is not clear, however, in what way attitude and achievement affect one another. This does not necessarily contradict the teacher's perception that more interesting and enjoyable work will lead to greater attainment. For one thing, the research does not deal with changes in achievement which might result within particular individuals or classes from improvements directed towards attitude; it rather shows that, broadly speaking, the set of people who like mathematics has only a relatively small overlap with the set of those who are good at it. However, research certainly suggests caution against over-optimism in assuming a very direct relation between attitude and achievement.

The discussion of the relationship between attitude and choice of mathematics for further study suggests that this choice depends more on the student's expressed liking for the subject than on its perceived easiness or usefulness. These factors are only weakly related to achievement: students who perform well in mathematics may still claim that they perceive it as a 'difficult' subject.

References for Chapter 9

AIKEN, L.R. (1976): 'Attitudes toward mathematics', *Rev Ed Res*, 46, 298-311.

ASSESSMENT OF PERFORMANCE UNIT (APU) (1980): *Mathematical Development: Primary Survey Report No.1.* London: HMSO.

CALLAHAN, W.J. (1971): 'Adolescent attitudes toward mathematics', *Maths Teacher*, 64, 751-5.

DUCKWORTH, D. and ENTWISTLE, N.J. (1974): 'Attitudes to school subjects: a repertory grid technique', *B J Ed Psych*, 44, 76-83.

ERLWANGER, S.H. (1973): 'Benny's conception of rules and answers in IPI mathematics', *J Child Math Beh,* 1,2, 7-26.

GOPAL RAO, G.S. (1968): The evaluation of certain intellectual and conative outcomes of modern and traditional mathematics teaching. Unpublished M.Phil. thesis, University of London.

GREENBLATT, E.L. (1962): 'An analysis of school subject preferences of elementary school children of the middle grades', *J Ed Res*, 55, 554-60.

HART, K. (1976): Mathematics achievement and attitudes of nine and ten-year-olds, effects of games and puzzles. Ed.D. thesis, Indiana.

INSKEEP, J. and ROWLAND, M. (1965): 'An analysis of school subject preferences of elementary school children of the middle grades: another look', *J Ed Res*, 58, 225-8.

JACKSON, P.W. (1968): *Life in Classrooms.* New York: Holt, Rinehart and Winston.

KEMPA, R.F. and McGOUGH, J.M. (1977): 'A study of attitudes towards mathematics in relation to selected student characteristics', *B J Ed Psych*, 47, 296-304.

KNAUPP, J. (1973): 'Are children's attitudes towards learning arithmetic really important?' *Sch Sci Math*, 73, 9-15, Jan 1973.

KYLES, I. and SUMNER, R.(1977): *Test of Attainment in Mathematics in Schools.* Windsor: NFER.

LEVINE, G. (1972): 'Attitudes of elementary school pupils and their parents toward mathematics and other subjects of instruction', *J Res Math Ed*, 3, 51-8.

LUMB, D. and CHILD (1976): 'Attitudes of teacher-training students to mathematicians.' *Ed Stud*, 2, March, 1-10.

NEALE, D.C., GILL, N. and TISMER, W. (1970): 'Relationships between attitudes toward school subjects and school achievement', *J Ed Res*, 63, 232-7.

PRESTON, M. (Undated): The measurement of affective behaviour in CSE mathematics. Research report, University of Leicester.

SELKIRK, J. (1972): *Report on an Enquiry into Attitude and Performance among Pupils Taking A-Level Mathematics*. University of Newcastle.

SHARPLES, D. (1969): Attitudes to junior school activities. *B J Ed Psych*, 39, 72-7.

STOODLEY, K.D.C. (1979): 'Attitudes of maths pupils to maths degrees', *Ed Res*, 21, 147-8.

Chapter Ten
Individual Differences

1.   INTRODUCTION

This section summarizes some characteristics which affect
the way in which individuals learn mathematics. The notion
being considered is that a particular child may respond
sensitively to certain experiences and learn from these while
other forms of experience may have little impact or indeed
may be actively resisted. The most effective learning situation is not the same for different individuals: the differences depend on intellectual and emotional factors, personality, and possibly sex and cultural background.

   Some research on individual differences attempts simply
to identify personality characteristics which affect mathematical achievement: more recent studies, however, have
tended to look for significant interaction between these
individual factors and teaching style, thus comparing the
relative effectiveness of different teaching approaches for
different children.

2.   PERSONALITY AND THE LEARNING OF MATHEMATICS

Consideration of personality characteristics and development receives scant attention in the extensive literature of mathematics education. The emphasis seems to be far more on cognitive issues - the development of schema, sequencing, problem-solving strategies, etc. Even where the personality aspects seem most relevant they often are ignored. Reports on attitudes to mathematics often deal solely with the material to be learnt and not with the learner. The literature on sex differences contains much discussion of the possible differences in abilities but very little about the differences in personalities. Problems on secondary school mathematics are discussed without any reference to the nature of the adolescent experience which the learners are undergoing.

This concentration on the cognitive is understandable in so far that with mathematics more than any other subject the pupil's work is likely to be judged in terms of simple 'right' and 'wrong' criteria and that repeated failure must play a major role in determining motivation and attitudes to mathematics. With other subjects, including science, there is usually a descriptive element, and pupils can gain credit for displaying partial understanding. Nevertheless most attempts to link academic success with cognitive abilities end up supporting the idea of a threshold, as first postulated by McKinnon (1962), such that pupils need a certain minimum cognitive ability to cope with a particular task but that many of those who possess that minimum ability lack the motivation to succeed.

It is reasonable to suggest that the learning of mathematics would be affected by personality considerations in a variety of ways other than just that of attitudes and motivation. The ability to think laterally or to break set may well be as much a personality characteristic, essentially a willingness to take risks, as a cognitive factor. When Hudson (1966) first studied divergent thinking among schoolboys he saw the characteristic essentially as being an ability which a pupil may or may not possess. But in his later work (Hudson, 1968), and as a consequence of considering further empirical evidence, he shifted the argument to that of seeing divergent thinking as being an ability most pupils possess but many choose not to use. This failure to think divergently apparently comes from a rule-obeying behaviour. The pupil assumes certain roles are applicable to the resolutions of the problem and conforms to these rules. That argument moves us close to the description of the authoritarian personality by Adorno *et al.* (1950) that such rule-obeying behaviour is characteristic of a distinct personality type which in turn is likely to be the product of particular childhood experiences. (Although views on the role of early childhood experiences have changed in recent years, see Clarke and Clarke (1976) and Rutter (1979), and the original Adorno work has attracted much criticism, e.g., Brown (1965), the main thesis developed above still stands.) In that event personality issues need to be considered in studying the learning of mathematics.

The current state of knowledge is such that any literature survey has to be brief. We can look for evidence from direct studies, e.g., those using psychometric tests of personality with students of mathematics, and from indirect sources, e.g., the work on attitudes and sex differences. Further insight may be acquired from examining the literature linking personality with the study of science in the belief that science and mathematics have enough in common to make comparisons useful.

3. DIRECT EVIDENCE FROM PERSONALITY TESTS

In the past two or three decades a number of psychometric tests of personality have been developed and used with student populations. The tests themselves are subject to much criticism. For example the Eysenck Personality Inventory, which is the most commonly used test of the type in Britain, only gives information on two orthogonal scales, extraversion-introversion and neuroticism-stability. Although the Eysencks have advanced many arguments for why these two factors are of paramount importance we might think that scores on just two scales represent an inadequate description of personality. Cattell's 16 Personality Factors test answers that criticism but in turn presents problems. Other workers e.g., Howarth (1976), have failed to detect all of Cattell's factors when they have analysed the test data. Consequently we need caution in approaching evidence from such tests. The failure to detect a set of personality characteristics associated with a particular subject choice, or success in a subject, may come from limitations of the tests employed rather than from the actual absence of such association, yet if statistically significant differences are to be found, say between students opting for mathematics and other subjects, then that result should provide some clues to work with in developing a model for linking personality with subject choice.

Oxtoby and Smith (1970) made a study of undergraduates entering the universities of Sussex and Essex in 1966. One of the measures employed was a test of tolerance towards ambiguity and uncertainty which revealed that the mathematicians tended to be about average in contrast to the arts students who were more tolerant and natural sciences who were less tolerant. A similar result was obtained with a measure of prejudice against racial and other minorities with the physical scientists being the most prejudiced. The importance of these results is that high levels of intolerance and prejudice are usually attributed to insecurity in the person holding such views. Head (1980) has suggested that the physical scientists, who are predominantly male, display the essential reticence characteristic of a particular

adolescent condition known as foreclosure. One way boys
can react to the emotional turbulence of adolescence is to
choose subjects with a tough emotion-free image. Science
meets that demand. Apparently that effect is absent, or is
much weaker, in those opting for mathematics.

Entwistle and Wilson (1977) made a full study of undergraduates from seven different universities. On many
characteristics mathematics students were close to the over-
all average, e.g., motivation, tendermindedness, social,
political and religious values. They tended to be rather
less extraverted and neurotic than most, with standard mean
differences (the difference in the means divided by the
weighted mean of the standard deviations) in the range
0.2 - 0.25. The introversion score was only matched by
students of history but the stability value was exceeded by
the physicists and nearly equalled by chemists. One problem
in interpreting such findings is that low scores on the
neuroticism scale may come from the responder deliberately
repressing the reporting of neurotic symptoms, an issue
long recognized by clinical psychologists, e.g., Blackburn
(1965).

Entwistle and Wilson also found mathematics students
to be somewhat conservative, with a standard mean difference
score of 0.18, which is not as great as that for physical
scientists and engineers, and to have the lowest aesthetic
values of any group (smd in the 0.52 - 0.56 range). Perhaps
the most important finding is that these students were far
and away the most syllabus-bound and the least syllabus-free
of all groups. On both measures the smd score was 0.47.
The characteristic of a syllabus-bound student is that he
resents departure from a prescribed syllabus while a
syllabus-free student welcomes opportunities to pursue
tangential issues which catch his interest. How might we
interpret this finding? More than other subjects, other
than perhaps philosophy, mathematics is self-contained, with
the tests of truth and validity being internal consistency,
in contrast with the natural and social sciences and history
which are concerned with describing events in the outside
world. Is that the feature being detected? It is nevertheless surprising that undergraduates should attach so much
value to the prescribed syllabus when clearly there is
considerable debate within mathematics education at all
levels about what content is best included in the syllabus.

One line of potential interest has been to look at the
association between personality and either success in or
liking for particular types of mathematics activity. A
study by Lewis and Ko (1973) examines the relationship
between performance in mathematics and two personality
dimensions, neuroticism-stability and extraversion-
introversion. This article includes a review of earlier
literature; and through this an underlying variation with

261

age may be observed, particularly in the effects of extraversion. A number of positive associations between extraversion and attainment are quoted for children up to the age of 12. Later in the secondary school this beneficial effect decreases, and instead introversion appears to take over as advantageous. With student samples from colleges and universities, introversion comes out strongly as an asset to academic performance. (See extensive list of references quoted by Lewis and Ko.) Rushton (1966, 1969) suggests that there is an age at which a changeover in the effect can be observed, and from his studies and those of Finlayson (1970) it would appear that this changeover comes within the age range 12-15. Lewis and Ko, however, in their work with fourth and fifth year secondary pupils, do not find a significant over-all effect in favour of introversion, although the superiority of introverts is to some extent demonstrated for girls: this finding may mean that the changeover occurs earlier with girls. Also observed here, however, is a significant extraversion x ability interaction. Among those in the higher level the introverts do better, the reverse applying among those of lower ability: this may indicate that the real determinant of when introversion becomes an advantage may be intellectual development rather than chronological age.

A study by Richards and Bolton (1971) considers whether divergent thinking contributes in a positive way to mathematical attainment. It is concluded that performance on mathematics tests 'is largely determined by a general ability factor, with the divergent thinking dimensions contributing to only certain of the mathematics tests and then only to a very small extent'. The authors suggest that the success of one school which used a combination of 'traditional' and 'discovery' methods lends weight to the suggestion 'that we must determine which kinds of information are learned most economically by authority and which by creative means.' This is a valid suggestion, but it is also plausible to argue that a variety of teaching methods can cater for a variety of individual children, and it may be appropriate to determine whether some *children* learn more effectively 'by authority' and others 'by creative means'.

Further direct evidence on personality factors is meagre. Krutetskii (1976) has a chapter entitled Mathematical Abilities and Personality which, in fact, says little about personality, in essence arguing that success comes from positive attitudes of enthusiasm and perseverance, which leaves open how such qualities develop. Begle (1979) finds nothing to report of interest and doubted whether research in this field was likely to be rewarding. Aiken (1976) reported that children who did well in mathematics tended to be conformists and obedient in school.

4.   APTITUDE-TREATMENT INTERACTION STUDIES

It may, of course, be that changes in the effects of
extraversion with age or ability are at least partly
attributable to the different teaching styles employed: it is
not difficult to believe for example that relatively informal
primary classrooms are more suited to extraverts than are
more tightly structured situations in the upper part of
secondary schools.   ATI (Aptitude Treatment Interaction)
studies have been used to determine whether certain teaching
approaches are more effective for different groups of
children.   Studies by Trown (1970) and Trown and Leith (1975)
have investigated the comparative effectiveness of exploratory
teaching methods and more 'traditional' instructional
methods with groups of primary and secondary school children.
The interactions between the difference in treatment and
certain personality dimensions (introvert/extravert and
anxious/non-anxious) are measured.   The significant interaction in primary school appears to be with anxiety and in
secondary school with extraversion.   Thus at primary level,
exploratory methods are less effective with anxious
children; but in secondary school anxiety becomes less
significant and it is the extravert children who gain most
from exploratory situations.   Other results (consistent with
these findings) have been reported of an interaction of
mode of learning with level of anxiety for 10 to 11 year old
children (Leith and Bossett 1967), and of an interaction
with extraversion-introversion for first year College of
Education students (Shadbolt and Leith 1967).

These results provide a kind of structure in which the
findings about the relationships between teaching style,
personality characteristics, age and mathematical achievement
can be understood.   They suggest that some children learn
better when they are 'free' to discover ideas for themselves,
but that others may not.   In primary school, for example,
a teacher might set up exploratory situations to provide
effective learning for non-anxious children, but at the same
time taking care to give anxious pupils the kind of support
and guidance associated with more didactive methods.

5.   FIELD DEPENDENCE

The most extensively studied difference in individual cognitive style is probably the field-dependence-independence
dimension: it is described in detail by Witkin et al. (1977).
Three tasks have been used to identify field-dependence.   In
the first, the subject (in a darkened room) is presented with
a tilted, luminous square frame, at the centre of which a
luminous rod is pivoted.   The task is to adjust the rod to a
position where it is perceived as upright, while the frame
remains in its initial position of tilt.   Marked individual
differences have been found, ranging from full alignment with
the frame (even up to 30° from vertical), to those who adjust

the rod close to the upright regardless of the position of the surround.

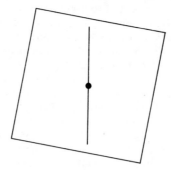

In another test, the position of the body is used: the subject sits in an adjustable chair in a small tilted room and is asked to adjust the chair to a position in which he experiences it as upright. Here again there are some people who can be tilted as much as 35°, and if that position is in alignment with the room, they will claim to be perfectly straight up ('this is the way I sit when I eat my dinner'): others appear able to disregard the surrounding room entirely. It may be noted that these differences do not occur if the task is conducted with eyes closed: the differences are the consequence of the conflict between the standard of the surrounding field and that derived from within the body.

In a third ('embedded-figures') test, the subject is shown the simple figure on the left; it is then removed and he is shown the complex figure on the right with the directive to locate the simple figure within it.

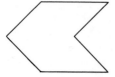 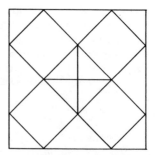

Again, for some people, the sought-after figure quickly emerges, whereas others are not able to identify the simple figure in the minute or so allowed for search.

In all three situations, the extent to which the surrounding field influences perception can be measured, and the results have been found to be consistent not only across these three tasks but also in auditory situations (where a

simple tune must be located in a complex piece of music) and
tactile ones (where, with eyes closed, the subject has to
pick out a particular pattern of raised contours). The
opposing modes of perception have been described as field-
dependence and field-independence: it may be noted that these
describe tendencies. Very few people are at either extreme,
and there is no implication that there exist two distinct
types of human being.

An awareness of this difference in cognitive style is
relevant in an immediate way for the teacher who seeks to
present ideas in various contexts, or to understand the
difficulties which some children have in identifying the
relevant features of a situation. The implications of the
distinction for choice of teaching style are less clear.
McLeod *et al.* (1978) suggest that field-independent students
will perform better when allowed to work independently, and
that field-dependent students will learn more when they have
extra guidance from the teacher: the results they quote
support this contention. A more recent study by Adams and
McLeod (1979), however, fails to confirm the expected
interaction between field dependence and level of guidance;
and in yet another experiment (McLeod and Briggs, 1980) in
which treatments differed in the use of an inductive or
deductive sequence of instruction (one dimension of discovery-
oriented teaching), this dimension is not shown to interact
strongly with field independence.

In summary, the notion of field independence provides
an explanation of why different individuals perform better on
some different tasks, and may suggest ways of improving
performance. Its implications have to be considered in terms
of specific tasks, and it does not give easy generalizations
about the appropriate choice of teaching style.

6. IMPULSIVE AND REFLECTIVE STYLES

The distinction between impulsive and reflective style has
been used by Bauersfeld (1972, 1976) in investigating
differences in performance in mathematics. The styles are
identified by means of a 'Matching Familiar Figures Test',
devised by Kagan (1970): the subject is shown a standard
picture (e.g., of a house with trees, fence, windows, a
chimney and so on), and then asked to choose from six
pictures *one* of which is identical to the standard. Twelve
such tasks are given, and the number of errors and response
times are noted.

		Response time	
		low	high
Number of Errors	high	IMPULSIVE	
	low		REFLECTIVE

By this procedure, it is suggested that something like 80 per cent of students can be fairly clearly identified as either impulsive or reflective: the dimension appears to be independent of sex and age. Bauersfeld's work shows, not surprisingly, that reflective students perform significantly better in arithmetic than impulsive students. Although significant, however, it is not a very large difference and explains only a small part of the variation in achievement in arithmetic.

Bauersfeld also refers to a study undertaken by Radatz (1976) which investigates the interaction between cognitive styles of teacher and student.

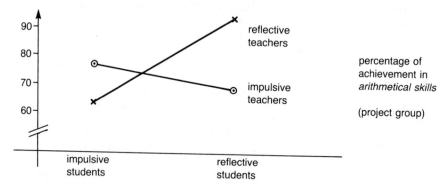

Matching of the traits has a positive effect: the reflective students, especially, seem to suffer considerably from mismatching. Radatz probes this a little more deeply, asking for impulsive teachers' assessments of highly reflective students: the comments quoted show underlying misinterpretations such as the following:

'... he does not say much, no reactions, seems to dream very often, - funny, by chance he comes out with a marvellous contribution, - his neighbour must have prompted him!'

Bauersfeld and Radatz have also studied this interaction in other topic areas: geometry, relations, solving word problems, and set language. The diagram overleaf shows the interaction for set language. Here the impulsive teachers do not do well with either type of pupil, and the impulsive students do not perform well with either type of teacher. The only significantly better achievement is that of the reflective students taught by reflective teachers. The nature of the interaction between teacher and pupil styles is to some extent content specific.

These ideas are summarized and clarified in a later article by Radatz (1979), in which the impulsive-reflective dimension is termed 'conceptual tempo'. The significant effect of this in arithmetic has already been noted: the

reflective pupils were also better in set language and
problem solving, but there was no significant difference
in geometry. Also in geometry the teacher-pupil trait
interaction while still positive was far less pronounced.

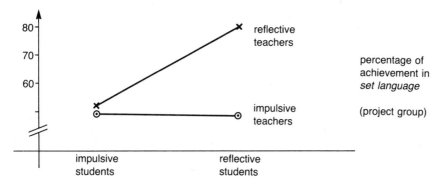

7.   OTHER ASPECTS OF COGNITIVE STYLE

A number of other dimensions of cognitive style have been
proposed. For example, an exploratory study of inductive
and deductive learning styles is reported by Gawronski (1972):
subjects are classified as inductive or deductive on one set
of tasks, but the results show little discernible difference
in performance for the two groups on further work. It may be
that the inductive-deductive tendency is subject-specific,
but this particular study does not support it as a general
dichotomy of learning style.

Pask (1976a) describes a distinction between 'holism'
and 'serialism', but these are put forward as alternative
learning-strategies rather than differences in individuals'
cognitive styles. Holism involves the consideration of
'many goals and working topics' at the same time, in an
attempt to achieve a global description or understanding:
serialism concentrates on 'one goal and working topic' at a
time, resulting in the mastery of topics in a step-by-step
fashion. In a later paper, Pask (1976b) suggests that some
students are disposed to holist strategies ('comprehension'
learners) and some to serialism ('operation' learners); but
there are also students able to act in either way according
to the nature of the learning outcome required - these
students are described as 'versatile'. Pask's studies do
not consider these differences in the context of mathematical
achievement; but the distinction is suggestive of a predis-
position to perceive learning either as a general structuring
of conceptual relationships or as the development of particular skills. If this is true, it indicates the need for
further investigation of the effects of this dimension,
particularly of the interaction of teacher and pupil traits
and of possible differing levels of performance in different
types of task.

Haylock (1978), investigating the relationship between divergent thinking in mathematical and non-mathematical situations, finds very little correlation between what he calls 'general' and 'mathematical' creativity: the latter is tested in geometrical and arithmetical situations. 'Creativity' measured in such mathematical situations apparently cannot be attributed to a simple combination of divergent thinking plus sufficient mathematical skills and knowledge.

8. STUDIES OF SEX DIFFERENCES

Currently about 40 per cent of the places in higher education are filled by women yet they only take 30 per cent of the mathematics places. The bias is even more evident with physics and chemistry, 12 per cent and 19 per cent respectively, and is most strong with engineering at 4 per cent.

Often the lack of women in mathematics and science is attributed to cognitive factors. Cognitive differences between the sexes have been extensively studied, e.g., Maccoby and Jacklin (1975) report on nearly 500 such studies in their annotated bibliography, and there is some consensus that males tend to do better with spatial and numerical problems and females with verbal tasks. The differences, however, are not great, most of the standard mean differences reported are in the 0.1 to 0.2 range, and many studies have failed to reveal significant cognitive differences at all. In that event Burton (1979) is probably correct in ascribing the lack of women in science more to affective factors. The subject is seen to be masculine so girls who possess the necessary ability will be put off from specializing in mathematics unless they have role models provided to help them make such a choice. Stamp (1979) has looked at sixth-form girls studying mathematics and found them to be more reserved, stable, tough-minded and radical than those specializing in French. That cluster of characteristics hints at the determination necessary for an adolescent to choose a subject against the sex stereotypes. Over and above the cognitive and sex-role issues Fennema (1979) suggests that girls face an additional problem in the responses of their teachers who tend to give more attention and specific criticism to boys.

9. SUMMARY

The research discussed here has identified a number of personality variables, cognitive styles and other differences between groups of children which affect achievement in mathematics. This effect is often related to the kind of material to be learned, the teaching method adopted and the teacher's own cognitive style. A recognition of these differences should at least provide an explanation of some differences in children's performance. It may not be practicable to structure the teaching of mathematics to take account of all the identifiable variables; but an awareness that persisting with a narrowly consistent teaching style may seriously handicap a proportion of children could lead to a better understanding of children's difficulties and indicate possible ways of remedying some of these.

References for Chapter 10

ADAMS, V. and McLEOD, D. (1979): 'The interaction of field dependence/independence and the level of guidance of mathematics instruction', *J Res Math Ed,* 10, 347-55.

ADORNO, T.W. *et al.* (1950): *The Authoritarian Personality.* New York: Harper and Row.

AIKEN, L. R. (1976): 'Update on attitudes and other affective variables in learning mathematics', *Rev Ed Res,* 46, 293-311.

BAUERSFELD, H. (1972): Einige Bemerkungen zum Frankfurter Projekt. In: *Materialen zum Mathematikunterricht der Grundschule.* Frankfurt, Arbeitskreis Grundschule.

BAUERSFELD, H. (1976): *Projects in Mathematics Education.* University of Trondheim Report, 87-100.

BEGLE, E.G. (1979): *Critical Variables in Mathematics Education.* Washington, DC.: Mathematics Association of America and the National Council of Teachers of Mathematics.

BLACKBURN, R. (1965): 'Denial-admission tendencies and the Maudsley Personality Inventory', *B J Soc Clinical Psych,* 4, 243.

BROWN, R. (1965): *Social Psychology.* Collier-Macmillan.

BURTON, G. (1979): 'Regardless of sex', *Maths Teacher,* April, 261-70.

CLARKE, A.M. and CLARKE, A.D.B. (1976): *Early Experience: Myth and Evidence.* London: Open Books.

ENTWISTLE, N.J. and WILSON, J.D. (1977): *Degrees of Excellence: the Academic Achievement Game.* Hodder and Stougton.

FENNEMA, E. (1979): 'Women and girls in mathematics - equity in mathematics education', *Ed Stud Math,* 10, 389-401.

FINLAYSON, D.S. (1970): 'A follow-up study of school achievement in relation to personality', *B J Ed Psych,* 40, 344-8.

GAWRONSKI, J.D. (1972): 'Inductive and deductive learning styles in junior high school mathematics: an exploratory study', *J Res Math Ed,* 3, 239-47.

HAYLOCK, D.W. (1978). 'An investigation into the relationship between divergent thinking in non-mathematical and mathematical situations'. *Maths in School,* 7, 2, p.25.

HEAD, J.O. (1980): 'A model to link personality characteristics to a preference for science', *Eur J Sc Ed,* 2,3.

HOWARTH, E. (1976): 'Were Gattell's "personality sphere" factors correctly identified at the first instance?' *B J Psych,* 67, 213-30.

HUDSON, L. (1966): *Contrary Imaginations: a Psychological Study of the English Schoolboy.* London: Methuen; Harmondsworth: Pelican Books.

HUDSON, L. (1968): *Frames of Mind.* London: Methuen.

KAGAN, J. (1970): 'Individuality and cognitive performance'. In: MUSSON (Ed): *Carmichael's Manual of Child Psychology.* New York: Wiley.

KRUTETSKII, V.A. (1976): *The Psychology of Mathematical Abilities in Schoolchildren.* University of Chicago Press.

LEITH, G.O.M. and BOSSETT, R. (1967): *Mode of Learning and Personality.* Research Report No. 14, National Centre for Programmed Learning, University of Birmingham.

LEWIS, D.G. and KO, P.S. (1973): 'Personality and performance in elementary mathematics with special reference to item type', *B J Ed Psych,* 43, 24-34.

MACCOBY, E.M. and JACKLIN, G.N. (1975): *The Psychology of Sex Differences.* Oxford University Press.

McKINNON, D.W. (1962): 'The nature and nurture of creative talent', *Am Psych,* 17, p.484.

McLEOD, D.B. *et al.* (1978): 'Cognitive style and mathematics learning: the interaction of field independence and instructional treatment in numeration systems'. *J Res Math Ed,* 9, 163 -74.

McLEOD, D.B. and BRIGGS, J.T. (1980): 'Interactions of field independence and general reasoning with inductive instruction in mathematics', *J Res Math Ed,* 11, 94-103.

OXTOBY, M. and SMITH, B.M. (1970): 'Students entering Sussex and Essex Universities in 1966: some similarities and differences, II', *Res Ed,* 3, 87-100.

PASK, G. (1976a): 'Conversational techniques in the study and practice of education', *B J Ed Psych,* 46, 12-25.

PASK, G. (1976b): 'Styles and strategies of learning', *B J Ed Psych,* 46, 128-48.

RADATZ, H. (1976): *Individuum und Mathematikunterricht,* Hanover: Schroedel.

RADATZ, H. (1979): 'Some aspects of individual differences in mathematics education', *J Res Math Ed*, 10, 359-63.

RICHARDS, P.N. and BOLTON, N. (1971): 'Type of mathematics teaching, mathematical ability and divergent thinking in junior school children', *B J Ed Psych*, 41, 32-7.

RUSHTON, J. (1966): 'The relationship between personality characteristics and scholastic success in 11 year old children', *B J Ed Psych*, 36, 178-84.

RUSHTON, J. (1969): A longitudinal study of the relationships between personality variables and some measures of academic attainment. Ph.D. thesis, University of Manchester.

RUTTER, M. (1979): 'Maternal deprivation, 1972-1978: new findings, new concepts, new approaches', *Child Dev*, 50, 283-305.

SHADBOLT, D.R. and LEITH, G.O.M. (1967): *A Further Study of Personality and Mode of Learning*. Research Abstract. National Centre for Programmed Learning, University of Birmingham.

STAMP, P. (1979): 'Girls and mathematics: parental variables', *B J Ed Psych*, 49, 39-50.

TROWN, E.A. (1970): 'Some evidence on the interaction between teaching strategy and personality', *B J Ed Psych,* 40, 209-11.

TROWN, E.A. and LEITH, G.O.M. (1975): 'Decision rules for teaching strategies in primary schools: personality-treatment interactions', *B J Ed Psych,* 45, 130-40.

WITKIN, H.A. *et al.* (1977): 'Field-dependent and field-independent cognitive styles and their educational implications'. *Rev Ed Res,* 47, 1-64.

Chapter Eleven
Language

1. INTRODUCTION

In the following discussion of language and mathematics education, various aspects of their relationship are identified. Firstly, the extent to which mathematical *concept formation is related to, and dependent upon language development* is considered: this includes the question of whether the development of adequate terminology is a prerequisite for cognitive growth. In a short second section, the *teacher's use of language* in mathematics teaching is discussed, and particularly how and why this varies from one class to another. Next, the use of *language in mathematical texts* is studied, followed by some comments on *language in set exercises, particularly in examinations*. The way in which *mathematics is expressed and written*, in particular, precise linguistic forms is then considered. Finally, some results about the learning of mathematics through *different languages* are quoted: consideration is given to the teaching of mathematics to immigrant children in their 'second' language, to bilingual children and to groups of mixed language.

2.   SURVEY PAPER - AUSTIN AND HOWSON

A survey paper in the field of language and mathematical
education, by Austin and Howson, was published in *Educational Studies in Mathematics* (1979): the paper began as an
attempt to compile an annotated bibliography, which the
authors then decided to preface by an essay. This essay is
not exactly a critical review of research in the field,
because it was felt that this would not adequately reflect
the way in which language and mathematical education
interact: rather, Austin and Howson's work indicates possible
areas for investigation and draws attention to areas where
research activities already exist.

The article is both thorough and extremely useful, and
the substantial bibliography is a good indication of the
wide variety of published material relevant to this topic.

The present chapter relies considerably on Austin and
Howson's paper as a source of references, and is an attempt
at a brief review of the present state of knowledge in the
field as revealed in written articles. This may mean the
proliferation of issues which seem peripheral and restricted:
but this in itself may reflect a general lack of awareness
of what the central issues are.

3.   LANGUAGE AND THE FORMATION OF MATHEMATICAL CONCEPTS

Piaget's earliest study of intellectual development (1926)
is entitled *'The Language and Thought of the Child'*: he
distinguishes between 'ego-centric' and 'socialized' talk.
In ego-centric talk, the child 'does not bother to whom he
is speaking nor whether he is being listened to'. It is
suggested that speech of this kind begins to disappear at
about the age of seven. In a later study (1954), Piaget
gives a more general view of the relationship between language and thinking, stating that

'... language and thought are linked in a genetic circle
where each necessarily leans on the other in independent
formation and continuous reciprocal action. In the last
analysis, both depend on intelligence itself, which ante-
dates language and is independent of it.'

Sinclair (1971) gives a useful summary of Piaget's
position on this matter. Vygotsky (1962) sees ego-centric
speech as a transitional stage from vocal to 'inner speech',
which is also called 'speech for oneself', the ability to
think in verbal terms. Austin and Howson note the 'sub-vocal
movements of the tongue and lips' sometimes observed when
difficult material is to be read and understood, and comment
on the tendency of mathematicians to want to talk to a
colleague in an apparently ego-centric manner in order to
analyse a difficulty. This use of language is clearly often

helpful: some consider it essential. Sapir (1963) writes:

> 'The feeling entertained by many that they can think, or
> even reason, without language, is an illusion ... no sooner
> do we try to put an image into conscious relation with
> another than we find ourselves slipping into a silent flow
> of words.'

There is, however, a lot of evidence of disagreement with
Sapir - for example, statements by Einstein and many others
(see Sheppard, 1978). Skemp (1971) quotes a number of
illustrations which, it is claimed, show the formation and
use of low order concepts without the use of language.

Nevertheless, language seems to play an essential role
in the development of higher order concepts. Both Piaget
and Vygotsky provide some evidence that the development of
linguistic structure in some cases precedes the appreciation
of the corresponding logical relationship. Piaget's experiments suggest, for example, that children use subordinate
clauses with 'because', 'unless', etc., long before they
grasp the logical relationship corresponding to these forms:
the statement 'grammar precedes logic' appears in both
Piaget and Vygotsky.

Vygotsky, in an experimental study of concept formation,
has probed more deeply into the dependence of the understanding of concepts on the language which describes them.
It is fairly obvious that children themselves do not make
the separation between linguistic form and meaning: the
identification of the linguistic forms as non-arbitrary
attributes of the objects they describe is seen by Vygotsky
as characterizing 'primitive linguistic consciousness'.

His simple experiments show that pre-school children
'explain' the names of objects by their attributes. Thus
when asked whether one could interchange the names of
objects, for instance call a cow 'ink' and ink 'cow',
children answer no, 'because ink is used for writing, and
the cow gives milk'. Language and thought are so inseparable
that names take the characteristic features with them in
the exchange. In one experiment, the children were told
that in a game a dog would be called 'cow': this is a sample
of the ensuing discussion:

'Does a cow have horns?'

'Yes'.

'But don't you remember that the cow is really a dog?
Come, now, does a dog have horns?'

'Sure, it is a cow, if it's called 'cow', it has horns.
That kind of dog has got to have little horns?'

Vygotsky's conclusion is not simply that concept and language are inextricably linked, but that concept formation depends on linguistic development:

'The birth of a new concept is invariably foreshadowed by a more or less strained or extended use of old linguistic materials; the concept does not attain to the individual and independent life until it has found a distinctive linguistic embodiment.'

The notion that concept formation arises through verbal discussion appears to have a fairly general acceptance. In his review of 'self-paced' mathematics instruction, Schoen (1976) deprecates the use of individualized learning schemes, stating that 'the educational quality of pupil-teacher interaction in the self-paced classroom is very poor, consisting mainly of procedural matters'. His explanation of the low effectiveness of such schemes is their failure to provide a 'quality of dialogue'. Stephens (1977), writing in similar vein, emphasizes the need for a varied pattern of communication. The learning of mathematics requires the 'negotiating' of mathematical meaning for (and by) each student: the use of prepared programmes places the teacher in too inactive a role for him to exercise this negotiation; and tends to isolate children from one another. Stephens expresses the view that the medium of instruction needs to be evaluated at least in part by its ability to develop relationships within which mathematical dialogue can be fostered.

An article by Hanley (1978) also discusses this area: his starting-point is some research on the process of thinking summarized by Wright and Taylor (1970). The role of language is seen as supplying verbal symbols which can represent concepts and be used as stimuli for the internalized manipulation of these concepts. There is some experimental evidence (Bourne, Ekstrand and Dominowsky, 1971, Ch.13) which indicates that problem solving is made easier by the provision of words to represent objects or ideas involved in the problem situation. Hanley's discussion has two main themes: firstly it is a plea that the oral verbalization of mathematical ideas should not be neglected, and secondly it provides some guidance about the choice of words for concepts, suggesting that these concept words should be applied carefully and not used loosely or ambiguously. An example especially likely to prove confusing appears to be the use of the words 'vertical' and 'horizontal' to develop the corresponding ideas. Hanley's analysis of this treatment in three different primary texts provides ample evidence of these words used in different ways and with meanings which change as the discussion proceeds.

While Hanley is critical of much use of language in texts to develop concepts ('the printed word has a permanence that demands accuracy'), he nevertheless allows (and expects)

276

the teacher to introduce ideas using words in an inexact way, which can be progressively refined until precision of thought is developed. He concludes that the best learning situations exist where language can be used freely as the interactive medium and the best resource for this is the teacher.

Much of this leads to the question of whether adequate terminology is a prerequisite for cognitive growth. Studies of attempts to teach mathematics without an adequate language structure appear to support this suggestion (cf. Philp in New Guinea, 1973, and Greenfield in Senegal, 1966). The problem of coping with deficiencies in vocabulary is an old one: Austin and Howson give several illustrations, and Easton (1966) describes Recorde's attempts in 1551 to encourage the use of 'lozenge' and 'threelike' instead of 'rhombus' and 'equilateral': it does after all seem reasonable to choose a word with familiar concrete connotations if no adequate desription of a concept already exists. Thus, according to Mmari (1975) there is no word used in Tanzania for 'centre': the Kiswahili word 'kitovu', meaning 'navel', has come to be used for the centre of a circle.

A particular conclusion here is that the development of terminology appears to be important for concept formation. A more general description of the place of language in the development of mathematical concepts is more difficult to provide, but the notion that language, and particularly oral dialogue, is an essential part of this process appears to be fairly well established.

4. ORAL LANGUAGE IN THE MATHEMATICS CLASSROOM

Austin and Howson (p.174) give a separate section to the way in which teachers use language: one aspect of this has already been considered in the discussion of the place of dialogue in concept formation, and particularly of the deterioration of classroom language when individualized schemes are used. There is clearly some kind of connection between the way a teacher talks in class and the relationship between teacher and pupils: this may not be specific to mathematics, but it is perhaps worth a brief mention.

American research, using 'interaction analysis' methods (Fletcher, 1960, 1970) provides some evidence that mathematics teachers talk more than social studies teachers, for example; they ask more convergent questions, make more directing statements, and elicit and reject fewer pupil responses. Various other differences between subjects can be identified: it is not clear whether these differences are inherent in mathematics or part of the background and tradition of teachers.

In Britain, attention has been focused more directly on the relation between language and teacher/pupil interaction. Henry (1971) describes 'a dialectic choice between mobilising children's attitudes to a pitch of excitement and attempting to control them'. The distinction made by Arnold (1973) is that in a teacher-directed class, children will learn to use (however uncomprehendingly) the teacher's language; whereas in a 'child-centred' class, unorthodox language presents the teacher with additional problems and new opportunities.

5.   LANGUAGE IN MATHEMATICAL TEXTS

A number of studies exist of the 'readability' of mathematics textbooks: generally the tests used were devised for analysing extensive prose passages but have been applied to mathematics texts. Aiken (1972) provides a summary of some of these investigations. A recent study of mathematical texts and of factors affecting their readability is reported by Rothery *et al.* (1980). An interesting feature of this study is its classification of categories of writing in textbooks, as follows:

E   *exposition* of concepts and methods, including explanations of vocabulary and notation, and rules.

I   *instruction* to the reader to write, or draw, or do.

X   *examples and exercises* for the reader to work on; usually these are 'routine' problems to be solved.

P   *peripheral writing* such as introductory remarks, summaries, meta-exposition (writing about the exposition), 'jollying the reader along', giving clues, etc.

S   *signals* - headings, letters, numbers, boxes, logos, etc.

These headings provide a crude system of analysis. Each heading represents a particular kind of response from the reader. The *exposition* is to be passively read and digested. *Instructions* require the reader to operate outside the context of the book and carry out the tasks described. *Examples* and *exercises* involve the solution of simple or complex problems; the reader must puzzle out what to do and how to do it, and then find the answer. *Peripheral writing* is to be read in a passive way though in contrast to exposition it need not require full attention and concentration; it is not likely to contain crucial information and it would not be important to try to remember the contents. *Signals* are of course not really 'read' though their use should convey a vast amount of important information; their purpose is not explicitly instructional; rather than teach Maths, they guide the reader through the page, helping identify and clarify the different parts of the text.

An example is given of a section of text analysed in

this way; although the technique is rather crude, it does concentrate attention on the purpose of the separate parts of the text and may lead to ideas for improvements:

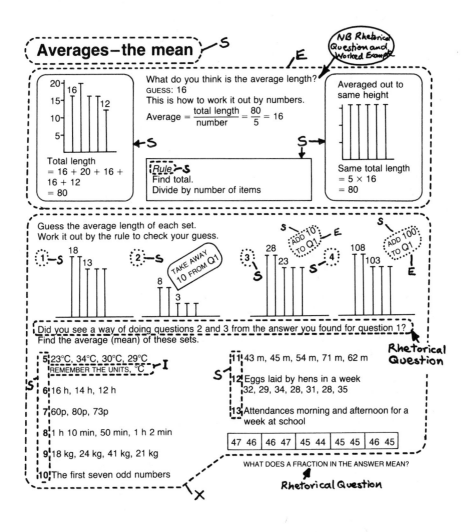

One of the identifiable changes in mathematics teaching over the past twenty years or so is the increasing demand on pupils to read and write in mathematics lessons. Texts are used not only as a source of examples, but as a medium of instruction; and the use of work-cards and independent learning has emphasized this even more. Increasingly, authors use easier words and less complicated linguistic forms (see Austin and Howson, p.173). This restriction of language is unsatisfactory in itself, and leads to the down-grading of word problems and a tendency to the 'verbal cue' situation of Nesher and Teubal (1975), already discussed in

Chapter Six. It is unlikely to contribute either to the
pupil's motivation or to his ability to apply mathematics.
But what alternative is there? Language difficulties may
hamper pupils whose mathematical development might otherwise
proceed apace. Call and Wiggin (1966) show that the provision
of special reading instruction can help to improve performance in the solution of word problems. It is arguable that
questions involving 'actual situations' should be read and
the situation discussed first, before pupils proceed to the
abstraction of mathematical relationships. Trueblood (1969)
suggests that illustrated problems, which pupils discuss
orally, permit development of both verbal skills and problem-
solving ability, and thus help to overcome reading difficulties. He also comments that interest is increased if such
problems are created by pupils themselves.

     In a general discussion of how to ameliorate language
problems in the use of textbooks, Rothery (1980) suggests
three lines of attack:

1.   Improve the text;
2.   Improve the teacher's use of the text;
3.   Improve the reading ability of the reader.

This section of the *'Children Reading Maths'* report is
perhaps the most valuable in making precise recommendations,
although its reliance on 'research' findings is a little
tenuous. Under (1), (improve the text), there is a
discussion of vocabulary problems, and then the following
points of prose style are given:

1.   Use short sentences;
2.   Use simple words,
3.   Remove unnecessary expository material;
4.   Keep to the present tense and particularly avoid the
     conditional mood. For instance 'If butter costs £2 a
     pack, how much would 5 packs cost?' can be replaced by
     'Butter costs £2 a pack. How much do 5 packs cost?'
     Words like 'if', 'suppose', 'given that' are traditional,
     though often unnecessary and confusing opening phrases;
5.   Avoid sentence structure which involves the reader having
     to remember clauses presented initially. For example:
       'Using a radius of 42 cm draw a circle ...' and
       'Taking π as 3.14 and assuming that the tank is a
       perfect cylinder, find the amount of water stored.'

Under (2), (improve the teacher's use of the text), some
suggestions are given for the usual situation where the
teacher is presented with pupils *and* a ready-made text which
may not be entirely suitable but which it is too late to
alter. Then under (3) a number of activities are suggested

which could help pupils to become stronger readers of
mathematical texts.

6.   LANGUAGE IN EXERCISES AND EXAMINATIONS

In this section, we consider how language is used and
interpreted in set exercises, especially in examinations.
The following is suggested as a model of the place of reading
and linguistic ability in ongoing mathematical activity (as
distinct from initial concept formation):

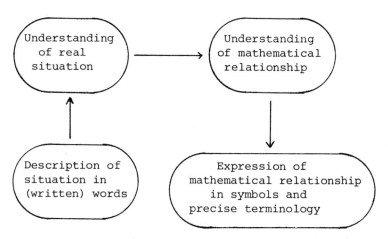

     This kind of process is needed to derive 'algebraic'
relationships from a word problem (especially when these
relationships are expressed in simultaneous or quadratic
equations) or, sometimes, to identify the appropriate
arithmetical procedure in a word problem. The left-hand side
is a process of interpretation, the top arrow is an abstraction (or, sometimes, modelling) process and the right-hand
side demands representation in symbols and special terminology. We have previously seen a number of examples in which
attempts to short-circuit the process across the bottom
produce incorrect results (Küchemann, 1977, Galvin and Bell,
1977) in algebra. Similar ideas in arithmetical problems
show up in the CSMS work on ratio, where some children look
for verbal cues in the wording rather than try to understand
the situation - thus 'percentage' for some always means
'divide by 100'. Nesher and Teubal (1975) have studied this
particular phenomenon, and show that reliance on such verbal
cues without over-all analysis of the situation is frequently
encouraged by an artificial mode of presentation of word-
problems in a specific and limited vocabulary. This is
reflected in both textbooks and examinations; so that certain
key words influence the choice of mathematical operation.
This must surely defeat the purpose of such word problems,
which are intended to encourage the revealing of underlying
mathematical relations in a situation, regardless of the
particular verbal formulation. What is going on here cannot

be merely a 'translation' from 'English' into 'mathematical symbols'. (Note that even translation from one language to another is not just 'verbal cues'; it demands first interpretation, then representation - there are plenty of ludicrous illustrations of the folly of word-for-word translation without understanding - but there is nothing to correspond with the modelling or abstraction stage in mathematics.)

In a discussion of examination questions, Graham (1977) considers the following (from East Midlands CSE, 1975):

(a)   In show-jumping a horse is penalised four faults for each fence it knocks down, and three faults for each refusal.
   (i)   If Manhattan knocks down x fences and refuses at y fences, write down an expression for the numbers of faults it incurs.
   (ii)   If Summertime finishes a round with 10 faults, how many fences did it knock down, and how many did it refuse?
(b)   In another show-jumping event, the horses jump two rounds, each consisting of seven fences. During the first round the horses score one point for clearing the first fence, two points for clearing the second, three points for the third, and so on up to seven points for the final fence. In the second round, seven points are subtracted from their first round total for knocking down the first fence, six points are subtracted for knocking down the second, and so on, down to one point subtracted if they knock down the final fence.
   (i)   Psalm jumps two clear rounds. How many points does it score?
   (ii)   Kingfisher knocks down the third and sixth fences on its first round, and the fifth and seventh fences on its second round. How many points does it score?
   (iii)   Sportsman scores 22 points in its first round. Which fences did it knock down? (There are four possible answers to this part of the question; give them all.)

There is scope for failure here in most of the ways just described: Graham discusses the difficulty under the heading of language, but the greatest difficulty seems to lie in the context more than anything else. In any case, the essential difficulties appear to have nothing overtly to do with mathematics. A number of examples have already been described where the abstraction process is made difficult by the unfamiliar context rather than the mathematical concepts (Boycott's batting average in the APU survey, and the 'snow-fall' problem in the NAEP survey). Since the context appears to account for a fair amount of variation in the facility of such tasks, some care is needed if these are intended to develop and test the understanding of mathematical concepts.

It must be pointed out that the arrows can go the other way: mathematical activity involves moving back and forth in this diagram. In this section, we have concentrated on problems in reading and understanding ordinary words (on the left): we now turn our attention to the difficulties of representation by and interpretation of symbols and terminology.

7. MATHEMATICAL TERMINOLOGY, SYMBOLISM AND LINGUISTIC STRUCTURE.

It is commonly stated that mathematics is itself a language. Obviously, the authors of this statement must consider that it possesses a certain degree of validity and has implications which make it worth saying. However, Love and Tahta (1977) regard this statement as a 'meaningless slogan' and Austin and Howson describe it as 'somewhat dangerous and potentially confusing'. There is an article by Sweet (1972) entitled *Children Need Talk*, a title which is in agreement with suggestions already made; and Sweet's discussion of concept formation in the article suggests that her experience of children's use of words in this development supports what has been said here. However, the article contains the misleading statement that 'as mathematics is a precise unambiguous language, our first concern must be to teach the language - then the grammar will follow'. It is probably best to go along with Austin and Howson's clear assertion that 'mathematics is *not* a language': it is an activity and a body of accumulated knowledge. It may in theory be enacted and described in a variety of languages, although an internationally recognized syntax and vocabulary has developed - but the system of symbols and terminology is not mathematics itself - any more than the words used to describe an experience are the actual experience. The language in which mathematics is presented is actually not so precise or consistent as might be popularly supposed, and Austin and Howson suggest the need for a mathematical counterpart to Fowler's 'Modern English Usage'. Of course, such a book would not solve the essential difficulties in understanding mathematical symbols and terminology. But beyond this, the way in which mathematics exploits the spatial properties of its symbolisms and develops 'manipulations' of symbolic expressions is a special characteristic which it does not share with ordinary language; it also is rather more prolific in its concept creation and hence its symbol creation than most disciplines. Some of these spatial aspects are discussed by Bell (1979).

It is obviously possible to study the use of symbols and terminology in books and examinations, just as the readability of ordinary language has been investigated - and it may prove more worthwhile. Graham's (1977) analysis of CSE questions includes the following (part question, East Anglian Board, 1975):

283

P, Q, R and S are subsets of $E = \{5, 6, 7, 8, 9, 10, 11, 12, 13, 14, 15\}$

$$P = \{x : x < 11\} \qquad R = \{z : z^2 < 80\}$$
$$Q = \{y : 9 < y \leqslant 14\} \quad S = \{\text{odd numbers}\}$$

(a) Place the sign ⊂ between the appropriate two sets.
(b) Which of the two sets are disjoint?
(c) Write the set $\{9, 10, 11, 12, 13, 14, 15\}$ in terms of R.
(d) List the elements of $Q \cap S$.

It is a question essentially about the use of symbols and words: the four sub-sets need to be first written out as lists of members. The answer expected for (a) is unique $(R \subset P)$ but of course any of the four sets could be given as a sub-set of E. (b) is virtually uninterpretable - presumably there is an error, but what kind of error? The request to write the set in (c) 'in terms of' R is a bit odd: it is of course the complement of R. Linguistically, the question is a bit of a mess. The earlier show-jumping question and this one are admittedly extremes, but they cast some doubt over what CSE mathematics examinations test. They may involve understanding of 'real-world' vocabulary and contexts, or of symbolic notation to which meaning has to be attached. Does either of these constitute 'mathematics'?

Some work on the analysing of mathematical terminology in primary texts and workcards has been carried out by Preston (1978). In eight schemes, he identifies 18 ways in which addition is presented:

3, 5 –   5 and 3 together make
5 and 3 equal   5 and 3
5 + 3   5 more than 3
5 plus 3   Add 5 to 3
3 greater than 5   5 count on 3
The total of 5 and 3 is   Write the number facts for 8
                                  using 5 and 3
5 add 3   What must be added to 5 to make 8
Add 3 and 5   Increase 5 by 3
The sum of 5 and 3 is   How many are there altogether?
                                  (Picture of 5 and 3 items)

One text, ostensibly designed for average/below average children, uses 14 of the 18 alternatives in two pages. Each of the eight texts uses, on average, seven alternative forms. It may be that the recognition of different forms is a worthwhile objective in mathematics, but it is certainly bad practice in terms of the standards relating to normal language development.

Otterburn and Nicholson (1976) have investigated the

understanding of words used in mathematics by 300 pupils
following CSE courses. The test instructions and table of
results are given below.

'On the left of the page is a list of words used in
mathematics. In column (1) put *Yes* if you understand what
the word means, *No* if not.
In column (2) put the *Symbol* for the word if it has one
(not all the words have them).
In column (3) *Draw a diagram* or use *Numbers* or *Symbols* to
show what the word means.
In column (4) *Describe in words* what the word means, use an
example if you like.'
An example was given:

	(1)	(2)	(3)	(4)
Word	Yes/No	Symbol	Draw a diagram	Describe in words
Plus	Yes	+	oo ooo oo + o = ooo oo oo ooo 4 + 5 = 9	Add, e.g. four plus five are nine

	Percentages		
	Correct	Blank	Confused
Minus	99.7	0.3	0
Multiply	99.7	0.3	0
Square	94	2	3
Remainder	92	8	1
Fraction	91	8	0.3
Rectangle	88	4	8
Parallel	77	19	3
Area	72	11	17
Decimal Fraction	68	29	2
Square Number	65	24	10
Radius	64	25	11
Perimeter	64	29	8
Volume	58	35	7
Prime Number	52	34	13
Kite	49	20	32
Reflection	45	51	4
Average	43	37	19
Intersection	41	50	9
Square Root	40	44	16
Rotation	37	60	3
Parallelogram	37	41	22
Perpendicular	35	61	4
Factor	32	62	6
Rhombus	31	47	22
Union	26	65	9
Ratio	25	71	4
Gradient	23	73	4
Symmetry	22	75	3

	Percentages		
	Correct	Blank	Confused
Product	21	59	20
Multiple	20	45	34
Similar	19	67	15
Congruent	18	77	5
Index	16	78	5
Mapping	16	81	3
Integer	15	76	9
Trapezium	11	79	10

(Note: the figures above are generally given to the nearest whole number.)

It is possible to question the importance of these words in understanding mathematics - but they are part of the language used in examination papers, where understanding of their meaning is presumably taken for granted. Again, what is being tested in such papers?

In a follow-up to this study, Nicholson (1977) has reported an investigation in which pupils and classes were tested so as to determine the extent to which they were handicapped by lack of mathematical vocabulary. It is of course difficult to isolate this as a critical factor, but it would be useful to know whether a considerable number of children who are able to understand the mathematical principles involved are seriously held back by lack of vocabulary. Nicholson suggests that, broadly speaking, the middle 50 per cent of the whole ability-range are significantly handicapped in this way. Sumner (1975) showed similarly the effect of low reading-ability on performance in written, as compared with practical, tests. In the classroom, it might be possible to devise and administer short, diagnostic tests to show up these difficulties; and some efforts might then be made to remedy them, perhaps on an individual or small-group basis.

Some of the difficulties arise not from the vocabulary of mathematical writings but from their linguistic structure. Austin and Howson point to the use of 'quantifiers' as a particular difficulty: quantifiers are often hidden in definite or indefinite articles, in words like 'always', 'everywhere' and 'sometimes' or are omitted entirely. Strevens (1974) also provides some examples, which show that many simple words and expressions (e.g., 'let', 'is' and 'for') are used in a particular linguistic form in mathematics to convey logical relationships. Thornton (1970) has shown the considerable confusion which this causes in mathematics, and the use of the word 'any' is especially shown to be associated with logical inconsistency. An earlier article by Thornton (1967) draws attention to the confusion which arises in manipulating symbols through use of learned phrases (cancelling, turning upside-down, taking out brackets, taking ... as, etc.); these are aids to automatic manipulation, but are not sufficiently informal to be distinguishable from the formal mathematical terminology. Interpretation and mathematical explanation of the processes would appear to be preferable.

8,   LEARNING MATHEMATICS IN A WEAKLY KNOWN LANGUAGE

A number of investigations exist of the effects of bilingual upbringing, or the change in language through migration, on children's mathematical upbringing.

     Austin and Howson report various such studies. Macnamara (1967) in his study of Irish children, and Morrison and McIntyre (1971) reporting on work in the Philippines and amongst Puerto Ricans in New York, suggest that 'bilingualism hampers a child's progress in problem solving but not in mechanical arithmetic'. Gallop and Kirkman (1972) have investigated the teaching and setting of examination papers (in both English and Welsh) in Wales and have reached the view that bilingual children may well be in an advantageous position over-all. Similar results are reported by Trevino (1968) amongst Spanish/English children, and by Giles (1969) in Canada.

     A survey paper by Engle (1975) on education through a second language provides a variety of material, not specifically mathematical. It appears to suggest, however, that there is no essential disadvantage (other circumstances being favourable) in starting to learn mathematics in a second language. It may be that some retardation may be attributable to weak understanding of the language being used; but this can eventually be made good with appropriate teaching and facilities.

     There is a sense in which this section might appear to contradict some earlier findings. Thus language factors have sometimes been shown to be very important: here language differences do not seem to have a significant effect. In all this, there is what may be termed a 'language variability question': on the one hand, there are those (such as Preston, 1978) who criticize the great variation in symbols and terminology; and on the other, there are situations in which expressing the mathematical ideas in more than one way may be beneficial. Situations in which the use of more than one system of notation may be either helpful or confusing, for example, could be quoted. The apparently advantageous effect of bilingualism may be a case in which mathematical abstraction is enhanced by linguistic variability. The way in which 'talk' is useful in mathematics teaching (in addition to the use of texts or work-cards) may also be an example of this: the written and verbal language together convey understanding more effectively than either on its own. Certainly the discussion by Watts (1979) of the particular problems of

287

teaching mathematics to deaf children, supports this idea.

References

AIKEN, L.R. (1972): *Language Factors in Learning Mathematics*. Mathematics Education Reports, Columbus, Ohio: ERIC.

ARNOLD, H. (1973): 'Why children talk: language in the primary classroom', *Ed for Teaching*.

AUSTIN, J.L. and HOWSON, A.G. (1979): 'Language and mathematical education.' *Ed Stud Math*, 10, 161-97.

BELL, A.W. (1979): 'The learning of process aspects of mathematics', *Ed Stud Math*, 10, 361-87.

BOURNE, L.E., EKSTRAND, B.R. and DOMINOWSKI, R.L. (1971): *The Psychology of Thinking*. Prentice Hall.

CALL, R.J. and WIGGIN, N.A. (1966): 'Reading and mathematics', *Maths Teacher*, 59, 149-57.

EASTON, J.B. (1966): 'A Tudor Euclid', *Scripta Mathematica*, 27, 339-55.

ENGLE, P.L. (1975): The use of vernacular languages in education. *Rev Ed Res*, 45, 2, 283-325.

FLANDERS, N.A. (1960): *Interaction Analysis in the Classroom*. University of Minnesota.

GALLOP, R. and KIRKMAN, D.F. (1972): 'An investigation into relative performances on a bilingual test paper in mechanical mathematics', *Ed Res*, 15, 1, 63-71.

GALVIN, W.P. and BELL, A.W. (1977): *Aspects of Difficulties in the Solution of Problems Involving the Formation of Equations*. Shell Centre for Mathematical Education, University of Nottingham.

GILES, W.H. (1969): 'Mathematics in bilingualism: a pragmatic approach', *ISA Bulletin*, 55, 19-26.

GRAHAM, J.D. (1977): 'Maths and CSE, Part 3', *Maths in School*.

GREENFIELD, P.M. (1966): 'On culture and conservation'. In: BRUNER, OLVER and GREENFIELD: *Studies in Cognitive Growth*. New York: Wiley.

HANLEY, A. (1978): 'Verbal mathematics', *Maths in School*.

HENRY, J. (1971): *Essays in Education*. Harmondsworth: Penguin.

KÜCHEMANN, D.E. (1978): 'Children's understanding of numerical variables', *Maths in School,* September.

LOVE, E. and TAHTA, D. (1977): 'Language across the curriculum: mathematics', *Maths Teaching,* 79, 48-9.

MACNAMARA, J. (1967): 'The effects of instruction in a weaker language', *J Soc Issues,* 23, 120-34.

MMARI, G.R.V. (1975): *Languages and the Teaching of Science and Mathematics with Special Reference to Africa (Accra workshop)*. London: Commonwealth Association for Science and Mathematics Education.

MORRISON, A. and McINTYRE, D. (1971): *Schools and Socialisation*. Harmondsworth: Pelican Books.

NESHER, P. and TEUBAL, E. (1975): 'Verbal Cues as an Interfering Factor in Verbal Problem Solving', *Ed Stud Math,* 6, 41-51.

NICHOLSON, A.R. (1977): 'Mathematics and language', *Maths in School,* 6, 5, 32-4.

OTTERBURN, M.K. and NICHOLSON, A.R. (1976): 'The language of (CSE) mathematics', *Maths in School,* 5, 5, 18-20.

PHILP, H. (1973): 'Article in Exeter Congress proceedings', *See* HOWSON, A.G. (Ed): *Developments in Mathematical Education.* Cambridge University Press.

PIAGET, J. (1926): *The Language and Thought of the Child.* London: Routledge and Kegan Paul.

PIAGET, J. (1954): 'Language and thought from the genetic point of view', *Acta Psych,* 10, 88-98.

PRESTON, M. (1978): 'The language of early mathematical experience', *Maths in School,* 7.

ROTHERY, A. *et al.* (1980): *Children Reading Maths.* Worcester College of Higher Education.

SAPIR, E. (1963): *Language.* London: Rupert Hart-Davies.

SCHOEN, H.L. (1976): 'Self-paced mathematics instruction: how effective has it been in secondary school?' *Arith Teacher.*

SHEPARD, R.N. (1978): 'Externalisation of mental images and the act of creation'. In. RANDHAWA, B.S. and COFFMEN, W.E. (Eds): *Visual Learning, Thinking and Communication*. New York, Academic Press.

SINCLAIR, H. (1971): Piaget's theory and language acquisition. In: ROSSKOPF, M.F., STEFFE, L.P. and TABACK, S.: *Piagetian Cognitive-Development Research and Mathematical Education*. Washington: NCTM.

SKEMP, R.R. (1971): *The Psychology of Learning Mathematics*. Harmondsworth: Penguin.

STEPHENS, M. (1977): 'Mathematics: medium and message', *Maths in School*, 6.

STREVENS, P. (1974): *What is Linguistics, and how may it help the Mathematics Teacher?* Paper for 1974 Nairobi Conference, UNESCO, Paris.

SUMNER, R. (1975): *Tests of Attainment of Mathematics in Schools: Monitoring Feasibility Study*. Windsor: NFER.

SWEET, A. (1972): 'Children need talk', *Maths Teaching*, 58, 40-3.

THORNTON, E.C.B. (1967): 'The power of words', *Maths Teaching*, 38, 6-7.

THORNTON, E.C.B. (1970): 'Any must go'. In: *Mathematical Reflections*. Cambridge University Press.

TREVINO, B.A.G. (1968): Ph.D. thesis, University of Texas.

TRUEBLOOD, C.R. (1969): 'Promoting problem-solving skills through non-verbal problems', *Arith Teacher*, 16, 7-9.

VYGOTSKY, L.S. (1962): *Thought and Language*. New York: MIT Press and John Wiley.

WATTS, W.J. (1979): 'Some problems in the teaching of mathematics to deaf children', *J Br Ass Teachers of the Deaf*, 3, 1, 2-6.

WRIGHT, D.S. and TAYLOR, A. (1970): *Introducing Psychology: An Experimental Approach*. Harmondsworth: Penguin.

Chapter Twelve
Evaluation and Assessment

1.   DEVELOPMENTS IN EXAMINING METHODS

The main change in the examining system over the last twenty
years has been the development of the Certificate of Secondary Education. One innovation brought by this was the control
of examinations by regional panels of teachers, rather than
of the GCE boards, and this, along with the attendant hierarchy of local district meetings to elect representatives,
has resulted in a certain kind of teacher control of the
examination. Possibly a more substantial innovation was the
provision for Mode 2 and Mode 3 examinations which brought
control of the syllabus, and, in the case of Mode 3, the
examining itself, into the hands of the teachers in the
school. These developments, though not only these, have
been responsible for an increase in interest in comparability
studies, which attempt to determine the extent to which
different syllabuses, different boards and different moderators are capable of determining a uniform standard in a
particular subject.

   Probably the most notable development in the technical
means of examining has been the trend towards the use of
short items, sometimes with multiple-choice answers, and
towards the assessment of course work as an element in the
examination. The advantage of short items is that the marking can be more efficient and reliable and may even be done

by a machine; it is also possible to cover a wider content; however, there is a danger that the questions will become more superficial, although this need not be so. Surprisingly, an investigation by Willmott and Nuttall (1975) indicates that increasing the number of items in this way has not led to CSE examinations being more reliable (in the sense of internal consistency) than the more traditional essay-type GCE examination.

A large number of GCE mathematics examinations have adopted short, objective questions on at least one of their papers. However, very few non-CSE mathematics examinations have contained an element of continuous assessment; one such was the London AO *Applicable Mathematics* scheme, recently modified, which was based on the development work of the Schools Council/ Reading University Sixth Form Curriculum Project. In this scheme, half the marks were awarded for project work in which mathematics is applied to some practical situation. Complex guidelines were provided for assessing this work, which took into account features such as the size, depth and significance of the investigation, the 'essay qualities' of the report, choice of mathematical model, mathematical accuracy, the level of generalization, the quality of their conclusions and their interpretation. The project assessment has been discontinued but a novel form of single-question paper remains.

A currently operating scheme is the AEB/Wyndham School O-level which allots 40 per cent of marks for a set of submitted mathematical investigations.

Here mention should also be made of an experiment conducted by the Scottish Certificate of Education Examination Board (1975) which investigated different ways of marking project work in sixth form physics. Previously low correlations between the assessment of the project work and a written paper led the board to consider a variety of marking schemes and moderation procedures for the project work, involving three categories of marker: the class teacher, one of the board's regular moderators (normally from a university), and neighbouring teacher who was himself submitting candidates. It was found that the most successful scheme was one in which the class and neighbouring teachers were required jointly to produce an assessment, coupled with over-all scrutiny by an external examiner.

Bloom's classification scheme

Another significant event of the mid-fifties was the publication by Bloom and his co-workers (1956) of Handbook 1 of their *'Taxonomy of Educational Objectives'* in which an attempt is made to identify the different kinds of intellectual process or behaviour that might be invoked in what children learn at school. The taxonomy consists of six major classes, shown below, which, it is claimed, form 'something' of a hierarchy, with 'Knowledge' (of 'specifics', of 'ways and means of dealing with specifics', of 'universals and abstractions', of 'theories and structures') representing the simplest behaviour and 'Evaluation' the most complex.

Knowledge
Comprehension
Application
Analysis
Synthesis
Evaluation

This classification scheme has come to dominate the way tests and examinations and, to a smaller extent, curricula, are analysed; the scheme is applied in practice by constructing a matrix with cognitive processes forming one dimension and subject content the other. Several attempts have been made to adapt the scheme to the needs of mathematics education (see, for example, Glenn, 1977, Chapter Six). Wood (1968) produces five categories, Hollands (1972) has nine and Avital and Shettleworth (1968) just three. The Mathematical Association (1976) lists 47 goals 'against which to measure or assess a mathematics course', of which 18 come under the heading 'Mathematical Knowledge and Intellectual Skills'.

The great strength of Bloom's analysis is that he has focused attention on the fact that mathematics (or any other subject) comprises many different mental activities, some of which may require different modes of teaching to be learnt most effectively. However, when the specific nature or his and similar taxonomies is examined, their value becomes questionable. Ormell (1974) argues that in many mathematics problems, the classes that Bloom tries to identify are inextricably bound together, and as such 'they cut across the natural grain of the subject, and to try to implement them - at least at the level of the upper school - is a continuous exercise in arbitrary choice' (ibid., p.7). Thus for example, with respect to 'Knowledge' and 'Application', Ormell makes the point that 'to know a theorem in mathematics and not to be able to apply it, is not to "know" it in any significant sense'.

A useful guide for groups of teachers in designing and constructing short-term tests of their desired objectives is Wilson (1970).

Evaluating general strategies and processes

Many teachers have become concerned with the fact that the low-level abilities which are easy to test using short items are by no means the only objectives for which they are working, and may not in the long term be even the most

important. One attempt to solve this problem has been the
development of course-work assessment; another has been the
attempt to devise means of evaluating the higher-level
strategies using written or oral items. These developments
are dealt with in the chapter on the learning of process
aspects of mathematics.

2. INTERVIEW METHODS OF EVALUATION

Since the production of the Nuffield Mathematics Project's
(1973) 'Checking Up' guides in the early 70's, a number of
local education authorities have begun producing guides for
teachers in primary schools (e.g., the ILEA 'Checkpoints')
which have the dual purpose of advising teachers on how to
assess their pupils' levels of understanding in a given
topic and of suggesting appropriate teaching materials in
the light of this assessment. The assessments are usually
designed to be carried out on an individual basis and often
involve materials that the teacher or child is asked to
manipulate. The guides give advice on the kinds of responses
to look for and ages at which these are expected to occur.

The emphasis on individual interviews for assessing
young children is understandable, and it is urged that all
teachers should find time to interview some of their pupils
in depth, whatever their age, if only to underline the fact
that children's thinking can be very different from that cf
an adult, and different in quite unexpected ways. However,
interviews are also extremely time-consuming, so that when
children are at an age to cope with written tests it is
worth considering these at least to supplement the interviews,
if suitable tests can be found.

It can be argued that compared to individual interviews,
the use of written tests results in both a distortion and a
loss of information: for example, it may be difficult to
determine the reasoning behind a given response, even when
the response is correct; or the child may give no response
at all because the item is too difficult, whereas during
an interview it would be possible to determine the degree
to which the item is understood. Usually it is also the
case that written tests are designed to describe the child's
performance in terms of a single score, which gives no indication of the kind of understanding the child has displayed
but at best ranks the child in relation to his peers on some
'over-all' mathematical ablility. Unfortunately this is the
case for most commercially available mathematics tests (see,
for example, the Mathematical Association's booklet 'Tests').
However, tests are now being developed that attempt to
identify the quality of children's understanding and which
provide guidelines for interpreting children's incorrect
as well as correct responses (e.g., the CSMS mathematics
tests which are being published by the NFER). Shayer (1978)
claims that once individual interviews have been used to

investigate thoroughly the different ways in which children cope with a given task, class tests can be developed (see Shayer *et al.*, 1980) that provide a more efficient and also more reliable assessment of children's understanding.

The mathematics section of the APU (1980) has taken the view that at least some mathematical activities ('practical mathematics') are best assessed individually. This is supported by an interesting feasibility study carried out by Sumner (1975), who compared children's performances on tasks that were presented in a practical and a written mode. For the practical items children were interviewed individually, whereas the written items were presented in the form of the class test. The written items tended to be more reliable (in the sense that children's responses were more consistent across items) and Sumner argues that the correlation of children's performances was too low to regard the two modes as equivalent. However, this is to some extent a matter of personal judgment and it may be argued that the written items did not adequately mirror the content of the practical items. Sumner also compared the effect of presenting a written test with the same test administered orally. As can be seen from the table below, the oral version was substantially easier for 11-year-olds with below-average reading ability, although the alternative forms of administration made little difference for the better readers.

Score groups of 11+ pupils above/below the median for Reading Comprehension

		(NFER Reading Test DE) RC score < 50%	RC score > 50%
Oral Test	Number of children	30	30
	Mean score	19.6	25.9
Written Test	Number of children	44	44
	Mean score	14.6	24.8
	Significance of difference between mean scores	1%	Not Significant

3. NATIONAL AND INTERNATIONAL SURVEYS OF ATTAINMENT IN MATHEMATICS AND OF CURRENT PRACTICE IN MATHEMATICS TEACHING

In 1964 the International Project for the Evaluation of Educational Achievement (IEA) undertook a comparative survey of attainment in mathematics of groups of secondary school children in 12 countries - 8 from Europe including England and Scotland, together with Australia, Israel, Japan and the United States (Husen, 1967). This was followed in the early seventies by surveys in six other subjects. Although the IEA's efforts have been widely praised in some quarters (eg. Bloom, 1974), the mathematics survey was criticized for

295

shortcomings (see, for example, Freudenthal, 1975) that cast some doubt on the value of its findings: for example, shortcomings in rationale (no mathematical educators were involved in the control and management of the project), organization (different countries administered the survey in different ways), test construction (no clear effort was made to take into account different countries' curricula) and research design (the variables that were thought might influence performance, across countries, such as 'size of school', 'teachers' salaries', 'level of mathematical instruction' were ill chosen, in that they often varied within as well as across countries, and in that they were often closely interrelated, but also often dependent on other critical factors that were specific to individual countries.

The table below shows some of the items given to samples of 13-year-olds in each country. Over all, Japan had the highest mean score and Sweden the lowest. The English sample had the greatest variability in scores, with a standard deviation twice that of Finland.

Test item	Average Difficulty	Order of error frequency	*Easier in*	*Harder in*
9. What is the square root of 12 × 75? A. 6.25 *B. 30 C. 87 D. 625 E. 900	.27	E, A	Austria, Belgium, England	Finland, Japan, Sweden
10. Three straight lines intersect as shown in the figure on the right. What is *x* equal to in degrees? A. 30 B. 50 C. 60 *D. 110 E. 150	.20	E [A, B, C]	England	—
3. In the division on the right, the correct answer is .004)24.56 A. .614 B. 6.14 C. 61.4 D. 614 * E. 6140	.58	B, D	Finland, France, the Netherlands	Australia, England, Scotland

A less ambitious study with interesting results was undertaken by Karplus et al. (1975). About 3,600 children aged somewhere between 13 and 15 years old were given a written proportional reasoning test (which included the well known 'Mr Short and Mr Tall' task) and a written control of variables test. The children came from a single town or region in the following seven countries: Denmark, Sweden, Italy, United States, Austria, Germany and Great Britain.

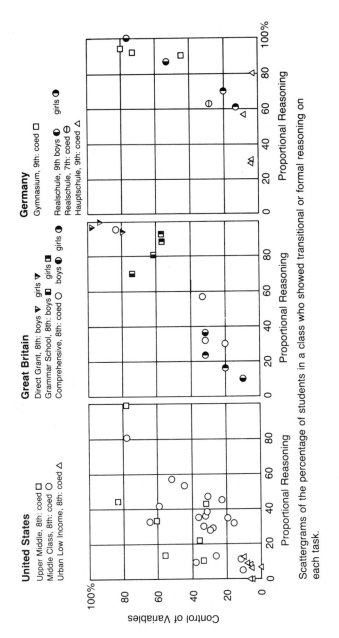

United States

Upper Middle, 8th: coed □
Middle Class, 8th: coed ○
Urban Low Income, 8th: coed △

Great Britain

Direct Grant, 8th: boys ▼   girls ▽
Grammar School, 8th: boys ■   girls ◨
Comprehensive, 8th: coed ○   boys ◑   girls ●

Germany

Gymnasium, 9th: coed □

Realschule, 9th boys ◑   girls ●
Realschule, 7th: coed ◓
Hauptschule, 9th: coed △

Scattergrams of the percentage of students in a class who showed transitional or formal reasoning on each task.

297

No claim is made that the samples from the different
countries are in any sense representative, but certain comparisons are still of interest, in particular the differences
in the relative difficulty of the two tests and the differences in the frequency of the 'addition strategy'. The
United States was the only country for which the control
variables test was generally easier than the proportional
reasoning test. For England and Sweden the tests were of
comparable difficulty, whereas the proportional reasoning
test was easier for the remaining countries, and substantially
easier for Germany and Austria (see above).

It can be argued that of the two tests, control of
variables is a more 'natural' test of reasoning in that this
particular experimental strategy is rarely taught explicitly
in schools. In contrast it is quite common for children to
be taught an algorithm for solving proportionality problems,
and it is possible that this is particularly prevalent in
Austria and Germany (or, to be more specific, Vienna and
Göttingen). Also, possibly, children from these countries
have come to rely more on taught procedures, so that they are
less used to coping with unfamiliar tasks. Some support for
this possibility comes from the fact that of the German
children who were unable to cope successfully with the
proportional reasoning tasks, very few resorted to the
untaught, but in many ways quite plausible addition strategy.
However, this was also generally true in Denmark, but not of
such children in Austria, who, together with similar children
in the United States and Great Britain, used the strategy most
often.

General surveys of practice in schools

In the United States, surveys by two national agencies, the
National Science Foundation (NSF) and the National Assessment
of Educational Progress (NAEP) are of particular interest.

The NSF, which has been heavily involved in the development of new curricula in science and mathematics, commissioned
three extensive studies of current practices and basic needs
in school science and mathematics teaching. The first of
these was a comprehensive review of developments in curriculum, instruction, evaluation and teacher education, undertaken, with respect to mathematics by Suydam and Osborne
(1977); the second was a series of surveys directed at
teachers, administrators, parents and pupils, directed by
Weiss (1978), and the third consisted of a series of case
studies undertaken in eleven school systems, together with a
small-scale survey, conducted by Stake and Easley (1978). A
useful summary of these studies is provided by Fey (1979a and
b). The over-all impression gained from this is that the
influence of new mathematics curricula on teaching methods
and content has been small: 'It appears that a large majority
of elementary teachers (still) believe that their sole

responsibility in mathematics teaching is to develop student
facility in arithmetic computation - this at a time when
availability of calculators has made such goals widely
questioned.' Secondary school mathematics teachers 'express
interest in learning new teaching ideas, yet seem basically
satisfied with their current methods and objectives' (Fey,
1979b, p.504).

Interviews of teachers as part of the case studies
provide graphic illustrations of these attitudes:

'I dislike our book, not enough drill, it's modern math.
We adopted a new book ... it has more drill, more basics
and I'll like it.'

'I am using the rote method pretty much because ... you
can spend all hour trying to get them to understand.'

'You can't hope to be creative until you've mastered the
basic program of studies.'

One of the observers commented,

'In all math classes I visited, the sequence of activities
was the same. First, answers were given for the previous
day's assignment. The more difficult problems were worked
by the teacher or a student at the chalkboard. A brief
explanation, sometimes none at all, was given of the new
material, and problems were assigned for the next day. The
remainder of the class was devoted to working on the
homework while the teacher moved about the room answering
questions. The most noticeable thing about math classes
was the repetition of this routine.'

Another observer in the same school system added,

'A general comment about math classes is that they were
dull. Science was perceived as fun ... but no one seemed
to look forward to math.'

Fey suggests a number of factors that seem to contribute
to this state of affairs, of which perhaps the most important
is the intense pressure that teachers are under, both from
within the classroom and from the influence exerted by
parents, administrators and society in general about what
teachers should be doing and achieving. With the growing
concern about 'falling standards' and the call to go 'back to
basics', these pressures have increased and the tendency to
avoid innovation reinforced. Teachers talk of 'going stale',
of 'burning out', and one observer describes this condition
as 'a flatness, a lack of vitality, a seeming lack of interest
in the curriculum by both the teacher and the children, a lack
of creativity and curricular risk-taking, a negativism toward
the children - they're spoiled, they don't care, they don't

try - and sometimes a negativism toward colleagues, administrators, and college and university training pro- gramme'. Probably the single most effective remedy for this condition would be to make more time available for in-service work, of a kind 'that would give teachers a chance to reflect, to recharge their knowledge and enthusiasm'. (Further comment on these studies is offered in Section C of this Review.)

Brophy (1979) provides an interesting review of American research into teacher-effectiveness (also further discussed in Section C). The review focuses on recent investigations of the teaching and organization in primary grade classrooms. Brophy identifies a number of clusters, or patterns, of teacher behaviour that are consistently related to learning gains (as measured by standardised achievement tests). One of these clusters is concerned with how efficiently the teaching is organized, and includes such variables as classroom management skills and the amount of time for which children are engaged on definite learning tasks. Another cluster suggests that 'students taught with a structured curriculum do better than those taught with individualized or discovery learning approaches, and those who received much instruction directly from the teacher do better than those expected to learn on their own or from one another' (ibid., p.4) (The type of discovery-learning referred to here is that in which children seek material on their own; discovery-oriented teacher-led discussion would be included in 'instruction from the teacher'.) Compared to older children, those in the early grades seem to require more recitation and drill, less genuine discussion, work that elicits few errors and that is pitched at a low cognitive level with very small and easy steps between one objective and the next. These results are based on large-scale correlational studies of classrooms, such as the *Beginning Teacher Evaluation Study* (McDonald and Elias, 1976) and the *Texas High School Study* (Brophy and Evertson, 1974; Evertson *et al.*, 1978). They suggest that, as seen in general practice, teacher-dominated learning environments were being operated more effectively than individualized methods. An observational study of English primary classrooms is currently in progress at Leicester University (see Galton and Simon, 1980).

The US National Assessment

The National Assessment of Educational Progress (NAEP) has completed two surveys of the mathematics achievement of 9, 13 and 17 year old students in the United States. The first was conducted in 1972-3 and the second five years later. Some items were common to both surveys, and these provide interest- ing information (NAEP, 1979) on the way achievement has changed during a period of growing demand that education go 'back to the basics': a period during which (according to NAEP, 1979) the use of text books emphasizing 'basic knowledge and skills' became more prevalent; during which schools were increasingly expected to be 'accountable' for pupils' performance, and during which a number of areas established required minimal levels of competence.

Over 70,000 children took part in the 1977-78 survey. In both surveys, between 2,100 and 2,500 children at each age responded to any one item, and the number of items common to both surveys was 55, 77 and 102 for the 9-, 13- and 17-year olds respectively.

The items were classified into four 'cognitive process levels': 'knowledge', 'skills', 'understanding' and 'application'. Looking at children's average performance on the items in each of the four categories suggests that mathematical achievement has declined in the five years since the first survey was undertaken. However, for the nine-year-olds the decline is slight, and it can be argued (NAEP, 1979, Ch.3) that the rather greater decline of the 17-year-olds may in part be due to a change in population, with more children staying on at school.

Increase, from 1972-3 to 1977-8, in Average Percentage of Children Answering Items Correctly

	Mathematical Knowledge Items (Exluding metric items)	Mathematical Skills Items	Mathematical Understanding Items	Mathematical Application Items	ALL items
Age 9	17 items -1	21 items 0	-	12 items -3	55 items -1
Age 13	15 items -2*	37 items -2*	12 items -2*	25 items -4*	77 items -2*
Age 17	16 items -2*	46 items -5*	13 items -4*	9 items -6	102 items -4*

*Change is significant at .05 level

Changes in performance on individual items are difficult to interpret. The view of a panel of consultants who were asked to comment on the results (NAEP, 1979, Chapter Three) seems to be that the changes reflect the changes in teaching brought on by the pressure to go 'back to basics', which, it is argued, 'has often resulted in a narrowing of the curriculum', with more attention focused on computational skills and knowledge of facts or definitions, and less time spent on problem solving' (page 25). There is support for this view, inasmuch as the greatest decline in performance on individual items came from amongst the Application items, and some of the

301

smallest declines (and in some cases gains) came from the Knowledge and Skills items.

However, over all there does not seem to be a clear relationship between changes in achievement and the change in teaching emphasis. Some Application items showed no decline at all, whereas some items that might be regarded as testing basics did. What this says about the validity of the surveys, the validity of the classification system, the claimed changes in the quality of teaching and the effect of teaching on achievement is an open question.

A selection of items that illustrate the ambiguity of the results is shown below.

Gain or Partial Gain

Knowledge				Skills				Understanding				Application			
	Age	1972–3	1977–8 increase		Age	1972–3	1977–8 increase		Age	1972–3	1977–8 increase		Age	1972–3	1977–8 increase
Order four 2-digit numbers	9 13	85% → 96% →	0 0	Multiply 2-digit by one-digit number	9 13 17	≈ 30% → ≈ 85% → ≈ 90% →	+ 4 0 0	What fractional part shaded	9 13	19% → 76% →	+ 1 + 5*	Word problem involving subtraction of 3-digit numbers, with regrouping	9 13	≈ 35% → ≈ 80% →	+ 4 + 3
Identify a decimal between two given decimals	13 17	≈ 15% → 46% →	0 − 7*	What is the sum of 21 and 54?	9 13	→ →	+ 6* − 9*	Carol earned D dollars during the week she spent C dollars for clothes and F dollars for food. Write an expression Using D, C and F that shows the number of dollars she had left.	13 17	≈ 17% → ≈ 46% →	0 0	Given 60 of each of these How many of these can be built?	9 13	3% → 24% →	+ 1 0
				Divide 3-digit by 2-digit number.	13 17	≈ 70% → ≈ 90% →	+ 4 − 3								
				43 71 75 + 92	9 13 17	≈ 50% → ≈ 80% → ≈ 90% →	− 3 + 1 − 2								

Decline

Knowledge				Skills				Understanding				Application			
	Age				Age				Age				Age		
Select a common proper fraction between two given common proper fractions	13 17	52% → 81% →	− 6 − 4	671 × 402	9 13 17	≈ 5% → ≈ 70% → ≈ 85% →	− 1 − 3 − 6*	If a 5 and b 5, what can you say about a and b? (Multiple choice)	13 17	67% → 83% →	− 10* − 7*	Word problem involving subtraction of 2-digit numbers, no regrouping	9	≈ 75% →	− 5*
				What does ⅔ of 9 equal?	9 13 17	18% → 56% → 81% →	− 5* − 7* − 8*	Problem dealing with transitive property of "is older than"	9 13 17	55% → 82% → 86% →	− 5 − 10 − 5	Word problem involving multiplication of 2-digit number, with regrouping	9	≈ 50% →	− 18*
				Express 9/100 as a percent	13 17	43% → 61% →	− 7* − 8*								

*Significant at .05 level

American and British children compared

The NAEP, APU and CSMS mathematics surveys have generated vast amounts of interesting data about children's attainment in mathematics. However, the restriction on the number of items made public by NAEP (1979 a,b;c,d) and APU (1980 a,b) means

Source	Item (or item-type)	Age (yrs):	9	11	13	15	17
NAEP	$\frac{4}{12} + \frac{3}{12}$				74		90
APU	$\frac{6}{10} + \frac{3}{10}$			69			
CSMS	$\frac{3}{8} + \frac{2}{8}$				66		
NAEP	$\frac{1}{2} + \frac{1}{3}$		1		33		66
CSMS	$\frac{1}{3} + \frac{1}{4}$				38		
NAEP	$-4 - +7$				21		39
CSMS	$-6 - +3$				30		
NAEP	$+7 \times -7$				36		65
CSMS	$+4 \times -4$				72		
NAEP	Find the area of a square, given the length of one side				12		42
APU	Find the area of this square 3↕□			37			
NAEP	Mike has 2 quarters. What is the probability he will get one head and one tail when he flips them? (Multiple choice, from $\frac{1}{4}$, $\frac{1}{3}$, $\frac{1}{2}$, $\frac{2}{3}$, $\frac{3}{4}$, don't know.)				60		69
APU	In a very large sample of women who have exactly two children, the probability of having a girl is the same as the probability of having a boy. What is the probability that a mother in this group has 1 girl and 1 boy? (Multiple choice, from $\frac{1}{4}$, $\frac{1}{3}$, $\frac{1}{2}$, $\frac{3}{4}$.)					70	
NAEP	Dale has 2 different slacks and 3 different shirts. How many different outfits (slacks and shirts) could Dale wear? (From 2, 3, 5, 6, I don't know.)		13		68		
CSMS	A shop makes sandwiches. You can choose from 3 sorts of bread and 6 sorts of fillings. How do you work out how many different sandwiches you could choose? (From 3×6, $6 + 3$, $18 \div 3$, 6×3, $6 - 3$, $3 - 6$, $6 \div 3$, $3 + 3$.)			46	72		

that it is difficult to compare the performance of American
and British schoolchildren. From the few comparable items
that exist, some of which are shown in the table on the
previous page, it would seem that the performance of American
and British children is quite similar. The clearest exception
is integer multiplication ($^+7 \times {}^-7$; $^+4 \times {}^-4$), though it is
interesting to note that for subtraction ($^-4 - {}^+7$; $^-6 - {}^+3$)
almost identical proportions of American and British 13-year-
olds (56 per cent and 51 per cent respectively) gave the
answer 3 (+ or -), which in each case would seem to be the
result of initially ignoring the sign of the given numerals.

4.   OTHER ASPECTS

Other aspects of evaluation are dealt with elsewhere in this
review. General evaluations of curricula in Britain and the
United States are treated in Section C, and the possibility
of evaluating pupils' general level of cognitive development
and relating this to the demand of the material in the
curriculum is discussed in the chapter on patterns of
development.

References

APU (1980a): *Mathematical Development: Primary Survey Report
No. 1.* London: HMSO.

APU (1980b): *Mathematical Development: Secondary Survey
Report No. 1.* London: HMSO.

AVITAL, S.M. and SHETTLEWORTH, S.J. (1968): *Objectives for
Mathematics Learning.* Bulletin No.3 of the Ontario Institute
for Studies in Education.

BLOOM, B.S. (1974): 'Implications of the IEA-studies for
curriculum and instruction', *Sch Sci Rev,* May, 413-35.

BLOOM, B.S., ENGLEHART, M.D., FURST, E.J., HILL, W.H. and
KRATHWOHL, D.R. (1956): *Taxonomy of Educational Objectives;
Handbook 1: Cognitive Domain.* London: Longmans.

BROPHY, J.E. (1979): *Advances in Teacher Effectiveness
Research.* Michigan: State University, Institute for Research
on Teaching.

BROPHY, J.E. and EVERTSON, C. (1974): *Process-Product
Correlations in the Texas Teacher Effectiveness Study: Final
Report.* Research Report No. 74-4. Austin: Texas: Research
and Development Center for Teacher Education, University of
Texas.

EVERTSON, C., ANDERSON, L. and BROPHY, J.E. (1978): *Texas Junior High School Study: Final Report of Process-Outcome Relationships, Volume 1*. Research Report No. 4061. Austin, Texas: Research and Development Center for Teacher Education, University of Texas.

FEY, J.T. (1979a): 'Mathematics teaching today: perspectives from three national surveys', *Arith Teacher*, 27, 10-14.

FEY, J.T. (1979b): 'Mathematics teaching today: perspectives from three national surveys', *Maths Teacher*, 72, 490-504.

FREUDENTHAL, H. (1975): 'Pupils' achievements internationally compared - the IEA', *Ed Stud Math*, 6, 127-86.

GALTON, M. and SIMON, B. (1980): 'Where the wild men aren't', *Times Ed Supp*, 4th April.

GLENN, J.A. (Ed) (1977): *Teaching Primary Mathematics: Strategy and Evaluation*. London: Harper and Row.

HOLLANDS, R. (1972): 'Aims and objectives in teaching mathematics', *Maths in School*, 1, 2, 3.

HUSEN, T. (Ed) (1977): *International Study of Achievement in Mathematics, Vol I and II*. London: Wiley.

KARPLUS, R., KARPLUS, E., FORMISANO, M. and PAULSEN, A. (1975): *Proportional Reasoning and Control of Variables in Seven Countries*. University of California, Berkeley.

MATHEMATICAL ASSOCIATION (1976): *Why, What and How*.

MATHEMATICAL ASSOCIATION (1978): *Tests*.

McDONALD, F. and ELIAS, P. (1976): *The Effects of Teacher Performance on Pupil Learning*. Beginning Teacher Evaluation Study: Phase II, final report: Vol. 1. Princeton, N.J.: Educational Testing service.

NAEP (1979a): *Changes in Mathematical Achievement, 1973-78*. Denver, Colerado: National Assessment of Educational Progress, Education Commission of the States.

NAEP (1979b): *Mathematical Knowledge and Skills*. Denver, Colorado: as above.

NAEP (1979c): *Mathematical Applications*. Denver, Colorado: as above.

NAEP (1979d): *Mathematical Understanding*. Denver, Colorado: as above.

NUFFIELD MATHEMATICS PROJECT (1973): *Checking Up III*. John
Murray and W.R. Chambers.

ORMELL, C.P. (1974): 'Bloom's taxonomy and the objectives of
education', *Ed Res,* 17, 3-18.

SCOTTISH CERTIFICATE OF EDUCATION EXAMINATION BOARD (1975):
*The Marking of Projects for the Certificate of Sixth Year
Studies in Physics*. Edinburgh: SCEEB.

SHAYER, M. (1978): A test of the validity of Piaget's
construct of Formal Operational Thinking. Ph.D. thesis,
University of London.

SHAYER, M., ADEY, P.S. and WYLAM, H. (1980): 'Group tests
of cognitive development: ideals and a realisation', *J Res
Sci Teaching*. (In press.)

STAKE, R.E. and EASLEY, J. (Eds) (1978): *Case Studies in
Science Education*. Urbana: University of Illinois.

SUMNER, R. (1975): *Tests of Attainment in Mathematics in
Schools*. Windsor: NFER.

SUYDAM, M.N. and OSBORNE, A. (1977): *The Status of Pre-College
Science, Mathematics, and Social Science Education: 1955-1975.
Volume II*. Columbus: The Ohio State University Centre for
Science and Mathematics Education.

WEISS, I. (1978): *Report of the 1977 National Survey of
Science, Mathematics and Social Studies Education*. Research
Triangle Park, N.C.: Research Triangle Institute.

WILLMOTT, A.W. and NUTTALL, D.L. (1975): *The reliability of
Examinations at 16+*. London: Macmillan Education.

WILSON, N. (1970): *Objective Tests and Mathematical Learning*.
Victoria: ACER.

WOOD, R. (1968): 'Objectives in the teaching of mathematics',
Ed Res, 10, 83-98.

Chapter Thirteen
Calculators and Computers

1. Calculators in the Classroom

2. Computers in Mathematics Education

1. CALCULATORS IN THE CLASSROOM

Personal electronic calculators are likely to produce
changes in the mathematics curriculum over a time-scale
significantly shorter than the decade or two which is normal.
The wide and rapid spread of ownership is already a fact in
most advanced countries and the natural inertia of educational
systems is insufficient for the calculator to be completely
ignored for long. The scale of curriculum changes to be
expected is unclear - caution and uncertainty as to how the
calculator should be used have ensured that so far it plays
a minor role in most classrooms and no role in many. However,
there are signs, as well as theoretical reasons, to suggest
that the impact may be substantial; if so, good research
results of various kinds have enormous potential value as a
guide to curriculum development - in pointing both to hopeful
directions and to pitfalls to be avoided in advance. When
curriculum change is partly 'forced' as in this case by
technological advances, any improved design or feedback
mechanisms that can speed up the process of effective curriculum development are valuable.

A list of desirable research was drawn up by a conference (National Institute of Education and National Science
Foundation) in 1976; it has recently been revised but the
sense of the recommendations remains. So far, some research
has been done and a fair number of PhD's awarded, but most of
it has done little more than clear some ground for the main
tasks. This summary will try to identify the essence of
what has been done, quoting some particularly interesting
examples, after classifying the main aspects of curriculum
development which it may aid. The Calculator Information
Center at the University of Ohio, directed by Marilyn Suydam,
publishes regular newsletters (1979, 1980) and reviews on
the absorbtion of calculators into the curriculum; these
reviews provide the basis of much that we shall say.

Suydam (1980) has also edited a collection of papers

prepared by the International Working Group on Calculators, reporting the place of calculators in schools in countries throughout the world. The picture is fairly consistent: calculators are in common use in employment, and in some countries almost every household has one, but many reports indicate that mathematics teachers may be reluctant to use them. Moreover, the resistance is greater among teachers of younger children; primary teachers are most resistant to their use, while university teachers allow them with almost no concern.

It is already clear that some curriculum changes are inevitable. Most people have expected as a minimum the disappearance of logarithms and other common tables as a means of calculation; though the time-scale of change for public examination boards is such that they currently remain in most 16+ examination syllabuses, sales of copies of the tables seem to have dropped by two orders of magnitude indicating acceptance of this change by schools and parents. Most observers expect very much more radical change than this, though none can tell its extent or nature.

We see curriculum changes of three different kinds, each with associated research needs:

(a)   the introduction of calculators with no planned modification of patterns of teaching and learning;

(b)   use of the calculator as a teaching aid as well as a computing aid with appropriately designed teaching procedures and materials, but with no modification of current curriculum objectives;

(c)   modification of the curriculum in the light of universal use of calculators.

So far the majority of studies have been on the basis of (a), usually seeking by pre- and post-testing of experimental and control groups to measure differences of attitude to or achievement on various kinds of mathematical activity. This sort of approach aims to provide statistically significant answers to some simple over-all questions; unfortunately the scale of the surveys almost always means that what actually happens in the classroom receives little attention and no control, except simply as to whether calculators are present or not. Not surprisingly, when so many elements of teacher variation are being averaged over, *most of the surveys find no statistically significant differences from calculator use* on most observed variables; *several authors,* however, *do find significant improvements* (in attitudes to mathematics, personal computational skills, understanding of concepts, and in problem solving), and *these findings tend to be in those experiments with better design,* particularly in control over classroom procedures. The general pattern, usually observed over a year or less, shows the familiar slow progress of pupils which seems to be independent of calculator use - this independence of the variable of interest is a familiar pattern in research on mathematical education and is, to labour the point, unsurprising when the learning environment is largely uncontrolled in experiment. *However, no negative effects appear in any study* - this is most important in view of the worries that have been expressed, particularly about the effects on personal calculating ability.

We shall comment specifically on a few of the results from well-designed experiments, before moving on to look at the more exploratory studies. Sharon Ayers (1977), in a carefully designed study of a college-level statistics course, found significantly better achievement from calculator use (the use of realistic situations in statistics was also investigated and led to improved student-attitudes). Billy Hopkins (1978), though finding improved computational skill only *with* the calculator, found clear evidence for the *improved learning of problem-solving skills* through calculator use; specially designed curriculum material was used for both with- and without-calculator groups in a well-designed experiment over 22 teaching days. Kasnic (1978), on the other hand, working over only nine days found no significant gain in problem-solving through calculator use, though it did *help low-ability students compete more successfully* with those of higher ability. An interesting study of proportional reasoning (Zepp, 1976) found no gain from calculator use, suggesting that a computational difficulty is not the major reason for the well-known switch to the erroneous addition strategy in more complex proportion problems with rational rather than integer scaling; however, only a single training session was involved so only the incurable optimist would be disappointed in these results. These studies are all from the USA.

A number of case studies have been reported with interesting results usually related to objective (b) above; as usual, they contain vastly greater amounts of information than the statistical survey but only the observers' impressions on the influence of the many variables discussed on the learning outcomes. The early work in Britain was largely concerned with the practicalities of calculator use, but generally contains some comments on curriculum possibilities with illustrative examples. The 1973-76 experiment in County Durham is probably the earliest example in this country, and a Schools Council Project has now been established there to develop material for using the calculator to help secondary school mathematics teaching. The 1973-76 project explored the uses in different parts of the curriculum: it became clear that in the theoretical parts of geometry and probability, for example, there was little direct use for the machine, but in the practical parts of the same subjects they came into their own. For example, the mensuration formulae within geometry and of course the whole

subject of statistics provide fields in which calculators are
almost indispensable. To take a particular example, the
theorem of Pythagoras, regarded as a statement about the areas
of squares of the sides of a right-angled triangle, is most
naturally treated theoretically, but as a statement about the
squares of the corresponding lengths, it offers a place for
calculator use. In the field of number, questions involving
ratio and proportion and best-buys in terms of unit cost were
made much easier by the machine. In theoretical number
topics, for example, Fibonacci series or recurring decimals
or number common multiples and highest common factors, the
calculator allowed a large number of examples to be easily
produced so that generalizations could be made and easily
checked. In algebra the study of functions and their graphs
was facilitated by the possibility of generating far more
points than was possible otherwise, and iterative methods of
solving equations became a more important part of the
curriculum. In general it appeared that less-able pupils
would be able to get much farther with genuinely applicable
mathematics than had previously been the case, instead of
spending a high proportion of their time on the mechanics of
calculation. In the sixth form the calculator was useful for
finding the sums of series, these and other limits, and for
numerical approaches to the solution of differential equations.

The SMP investigation (1976-77) similarly led to a
report which commented in a preliminary way on a wide range of
issues. In both cases, separate reports from the individual
schools involved in the experiments are given, providing
interesting evidence on the range of approaches and attitudes
among the pioneering teachers involved. (The full pattern of
curriculum development following on the SMP study is unclear.
The role of the calculator in the current complete revision
of the main SMP material for the 11-16 age-range presents the
project with difficult decisions; it will be interesting to
see what emerges. Meanwhile, SMP are playing a pioneering
role in the development of an O-level examination component
with calculators as an integral part; this is discussed below.
The other major curriculum effort known to us is by Peter
Kaner, who is writing for Bell and Hyman a five-year 11 - 16
mathematics course at three ability levels in which a
calculator is assumed to be an integral part of each child's
equipment.)

There have been a few attempts to look in more detail at
the possibilities for calculator use in mathematics teaching.
Lowerre, Veneski and the Scanduras (1976), in a series of
reports over several years on teaching experiments with a
few children in the age range 5-9, explored the teaching of
various parts of the primary curriculum with calculators as an
aid and identified hopeful areas. The very special nature of
the teaching environment and the impressionistic nature of
the observations means that these results should be taken as

indicative rather than conclusive. However, they are of interest to all other researchers and curriculum developers in this area. Bell, Burkhardt, McIntosh and Moore (1979) made a detailed observational experiment in a British primary school, systematically identifying both different ways of using the calculator and parts of the curriculum where it appears to help. No objective measures of relative pupil-progress were attempted; the aim of this work was to go on to the current phase of developing calculator-aware teaching material which can be improved and absorbed before summative evaluation is attempted.

The total effort in research and curriculum development in the UK seems surprisingly small. The Schools Calculators Working Party, which is an informal group for those engaged in experimental work in the use of calculators in teaching, has made contact with all groups known to be engaged in moderately systematic work in either area, but their numbers remain small.

In view of the importance of the examination system in determining secondary curricula, changes in the role of calculators in that area are important. The situation is fluid but it remains true that the GCE boards are taking steps to include the calculator as an integral part of the curricula under their influence, while the CSE boards are generally moving slowly and with apparent reluctance from a position of total rejection. CSE attitudes have been stated primarily as a concern for the importance of personal basic arithmetic skills in all subject areas, and for ensuring fair and equal treatment of all candidates in an age where calculator ownership is not universal. While research and time together offer means of allaying both these fears, the inertia of the system should not be underrated. The GCE boards have chosen different ways of absorbing the calculator into their syllabuses. The largest of them the JMB, took a decision in 1978 that calculator-use should be freely permitted in all their examinations, but that questions should be set in such a way as to confer no significant benefit on those who used calculators over those who did not.

This decision severely limits, of course, the exploit-ation of the calculator's potential, and it is to be hoped that the restriction is a temporary one, allowing time for the basic decision to encourage universal calculator use in the Board's examinations. Certainly a most interesting research study by Peter Connah (1979) suggests that calculator-use does in fact confer significant benefit on candidates, even when the questions may have been designed to avoid this. In contrast, the SMP O-level mathematics examination is now available in two versions, one of which forbids and the other of which requires calculators to be used. A common paper without calculators serves as a moderating instrument. Since public examination papers epitomize the curriculum objectives in secondary schools, this approach allows their modification, and the evidence of the first two papers, set in 1979, shows this to be proceeding, albeit cautiously. This examination was supported by a series of booklets on calculator-use in particular topic areas, published by SMP (1979); together with the papers and the Awarders' Report (1979), they present a picture of the process of gradual curriculum reform, with objective feedback from candidates' scripts, which characterize the forward-looking operation of the public examinations boards.

There is, of course, a fair amount of work in progress in other countries; it includes a most interesting study in Sweden of the effects of introducing a calculator-based curriculum, and is thus in category (c) above. All secondary pupils in Sweden have been provided with calculators. In ten schools, the classes of 10-year-old pupils are studying a curriculum in which calculators are continuously available - the emphasis on personal computation is much reduced with more time being devoted to learning problem-solving of various sorts. A test has been developed and standardized to measure progress towards a definite set of higher level objectives and the performance of the experimental and control groups will be compared. An interesting feature of the work is the provisional funding of an intense programme of remedial teaching of arithmetic ('the Fire Brigade') which will be implemented if the experimental group is shown to have been disadvantaged; this device may be a useful way of allaying public doubts about the ethics of experimental teaching programmes (no corresponding action will be taken on behalf of the control group if they are shown to have been disadvantaged!). This work was described at the May 1978 ICME meeting on Calculators in Secondary Schools: it has not yet been published. Other work in progress includes both observational studies and more controlled experiments of the types reported here; it is complemented by the production and informal testing of a fair amount of teaching material. In both areas, however, we are still right at the beginning of the reform process.

Finally, an anecdotal note from South Australia, where such changes have been steadily spreading since 1975. In the High Schools, which start at the age of 12, calculators are used except where specifically excluded in the teaching of certain topics. They are mostly owned by the students, but there are also calculators scattered around the mathematics areas, and they are available from the library on a book-loan basis. The curriculum is gradually changing - logarithms are not used for calculation, and calculators are explicitly built into trigonometry teaching. Decimals appear earlier, while the study of fractions is confined to small denominators with less emphasis on their arithmetic. There is more work on estimation. Calculators are available in examinations.

2. COMPUTERS IN SCHOOL MATHEMATICS

Previous curriculum development efforts with computers in
school mathematics can generally be categorized under two
headings. The first is tutorial computing, including drill
and practice programs, and the second is best described as
programming for concept demonstration, concept reinforcement
and 'problem-solving' (problem-solving is the description
most commonly given to this work, but it does not give a
full picture of the very varied emphasis in many of the
projects). More recently, demonstration programs have been
developed combining both aspects, so that the teacher and the
microcomputer are working in partnership.

We will restrict our remarks mainly to the actual use
of the computer in mathematics classes and particularly to
the use of the computer by teachers and pupils as a tool for
investigating mathematical relationships and problem-solving.
There have been, and are currently, many interesting tutorial
computing activities in mathematics (see for example NDPCAL,
Final Report of the Director, Richard Hooper, Council for
Educational Technology, 1977). However, this type of use is
not unique to mathematics and while the results of such
activity are encouraging and may well reflect a major school
use in the future, there are many problems in large-scale
implementation of such activity, which are unlikely to be
resolved in the next few years. In particular, such use
usually requires considerable time and effort to produce a
limited, small, computer-based teaching programme. In
addition, the present cost of adequate equipment (terminals,
or microcomputers) is beyond the budget of most schools since
the approach generally assumes a one-to-one interaction
between pupil and computer, and hence the need for many
input-output devices.

Within the restricted domain of computer investigations
in secondary school mathematics the main characteristic of
most efforts in this area to date, and that which differ-
entiates this use from activities in other disciplines, is
that pupils write their own programs, though among important
exceptions to this rule is the use of stored programs for
computing parameters and performing analyses in the field of
statistics. Some examples of curriculum materials which
emphasise a programming approach are:

Computer Assisted Mathematics Program (CAMP), directed by
D.C. Johnson, 1969-72.

Project SOLO, directed by T. Dwyer, 1973.

*SMP Computing in Mathematics: Some Experimental Ideas for
Teachers,* edited by J.D. Tinsley and B.H. Blakeley, 1971.

SMP Computer Oriented Mathematics Package (COMPACK), edited
by Rosemary Fraser, 1975.

The 1971 SMP book is included to provide an impression of early efforts in mathematical computing in the UK. However this book differs from the other projects in that no computing or programming language was assumed and a great deal of the material could be implemented without actually using the machine - the emphasis was on the nature of algorithms and iterative processes presented in the form of flow diagrams. However, all these efforts reflect to some degree the notion that pupils could write programs to investigate mathematical relationships and that the computer program or program output would serve to

perform tedious, repetitious calculations;
demonstrate and reinforce selected concepts;
assist in the investigation of scientific or mathematical phenomena and the problems they suggest.

The two US efforts, CAMP and Project SOLO, also included research components. The CAMP project was based at the University of Minnesota High School, a laboratory school associated with the University; it worked mainly with grades 7-12 (ages 12-17). The project developed curriculum materials, student texts published by Scott Foresman and Co., Glenview, Illinois, USA, for use as a supplement to the regular textbook at each grade level. The five textbooks are:

Hatfield, Johnson et al., CAMP First Course (age 12)
Walthes, Johnson et al., CAMP Second Course (age 13)
LaFrenz, Johnson et al., CAMP Algebra (age 14)
Katzman, Johnson et al., CAMP Geometry (age 15)
Kieren, Johnson et al., CAMP Intermediate Mathematics
(age 16 and 17)

They are not currently readily available in the UK. Each of the books has an introductory section on programming in BASIC and the subsequent chapters identify settings in the curriculum which could be enhanced with computing. This approach is also illustrated in the SMP COMPACK workcards. In addition to the development of materials the CAMP team conducted a number of experiments at different levels. The emphasis was on studying the contribution of the programming activity to the understanding of selected concepts in the standard curriculum of the time. The CAMP research was also supplemented by a number of related studies undertaken by graduate students and staff. The over-all findings gave evidence that supports the contention that the writing of programs to study selected concepts enhances learning - favourable results were reported for the study of number theory, series and sequences, functions and limits, among other topics. Some of the most exciting research results to date are in the area of problem-solving, including the work of Foster (1973), Anderson (1977), and Johnson and Harding (1979); this last UK study has produced evidence to the effect that university students who study mathematics with computing are more apt to employ different problem-solving strategies effectively and perform significantly better on mathematical problem-solving tasks.

Project SOLO was a joint effort of the University of Pittsburgh and the Pittsburgh Public School System. The Project SOLO materials cover a wide range of mathematical and scientific topics and the mathematical areas are described in a series of booklets under the general title *Project SOLO Computer Topics: Mathematics Projects*, published by Hewlett Packard (Cupertino, California, USA). Basically the materials identify a number of projects (problems) which are appropriate for a given level. The Project SOLO reports are primarily descriptive narratives of the types of projects completed by pupils. The reports provide case study data which indicates that pupils are able to attack a wide variety of interesting problems which involve the application and/or the extension of the mathematics they have been studying.

These past efforts represent one major aspect of how and why pupils should use the computer in their study of mathematics, namely that one can better teach selected aspects or components of the present curriculum. However, there are other points which should also be noted. For example, the projects certainly show that it is now possible to study topics which prior to computer availability were difficult, if not impossible on practical grounds, to include in the curriculum. Further the importance of the study of processes, procedures and algorithms, becomes a major topic of interest, particularly when one considers that tomorrow's user of mathematics will undoubtedly have access to and be expected to use a computer. In some areas today's user is already in this position.

It would be useful to go on and consider how computing might influence and change school mathematics. However, such a discussion goes beyond the scope of a review; suffice to say that the influence is likely to be considerable, with more emphasis being given to such topics as probability and statistics, numerical algorithms and algorithm design, and mathematical modelling.

To return to the current situation, in the US some mathematics textbook series are incorporating aspects of the early computing projects, but this is not common in the UK. The usual approach is to include a short unit on programming and then provide supplementary or enrichment problems at various points in the text. However, even with this approach the amount of computer use implied is still quite small. Hence one might be tempted to ask why, since *the results from projects in the 60s and 70s are so positive and there are materials readily available*, there isn't more computer programming in school mathematics courses? The materials

produced by projects in the US and the UK have received
wide distribution and, in general, very positive feedback
from teachers. Further, the case for computing in mathematics is continually being made in the professional journals
(see for example, Hart and Elliott, *Mathematics in Schools*,
March 1979, and Peele, *The Mathematics Teacher*, February,
1979). The answer is surely that such curriculum reforms
spread only very slowly, unless there is strong social
pressure, and, in this case, full hardware availability.
In addition, it is clear that the implementation of such an
approach requires expertise, both knowledge and training, on
the part of the teacher, which includes skill in programming
as well as familiarity with the hardware. All this requires
no small effort on the part of an already busy person.
Further the activities, particularly in problem-solving,
take time and as such often seem to impose additional
problems for the teacher who feels pressure to 'cover' the
established syllabus. Finally, there is the very real
problem of classroom management and, in particular, the
management of equipment resources. It is extremely difficult
to process the many programs prepared by individuals, or
even small groups, with the limited resources available in
most schools - often a single terminal or micro-computer
for which there are also demands on use by other disciplines.

Despite these restrictions, successful programs have
been launched in a number of areas of the country, particularly Inner London, Birmingham and Hertfordshire.
Hertfordshire has produced a substantial computer-managed
mathematics curriculum package for use in its schools which
provides the teacher who chooses to use it with many aids,
including the automatic marking of tests by computer. The
full use of the facilities of the package, however, inevit-
ably restricts the freedom of the teacher in determining
the curriculum. The London Borough of Havering has also
extensive experience in computer-managed learning which has
also been explored in Birmingham. These systems are established but still to be regarded very much as experimental;
their maintenance requires great efforts by the individuals
concerned in meeting the multitude of practical problems
that arise. They nonetheless represent a substantial
achievement in pioneering enterprises.

The rapid drop in price of computing power that has
come in the last two years with the spread of microcomputers
has introduced a new phase of development. Many schools are
now equipped with a microcomputer and visual display units,
and often this equipment is housed in the mathematics
department. Programs are being designed to use the microcomputer in demonstration mode as an aid to the teacher in
the classroom and it is beginning to emerge that the single
microcomputer used in this way can become a powerful teaching
aid. Two projects which are currently investigating this

approach in mathematics, as well as in other subjects, are:

Schools Council Computers in the Curriculum Project,
Birmingham Educational Computing Centre;
Investigations on Teaching with Microcomputers as an Aid
(ITMA), College of St Mark and St John, Plymouth.

The work is all at an early stage of development. A few
units have been developed that show promise while revealing
some potential problems in program design if they are to be
used by a wide range of teachers. The Schools Council work
has concentrated on exploring a wide range of potential
techniques, particularly in the use of graphics, while the
ITMA team has placed more emphasis on tackling specific
curriculum problems. The systematic approach to classroom
development, which the ITMA team is developing with the
Shell Centre for Mathematical Education, is backed by an
SRC-funded basic research project which is studying the impact of the teaching microcomputer on the patterns of interaction of teacher and pupils in the classroom as a case
study on human interactions with computers in already complex
situations. Particular interest centres on the ways teaching
objectives may be modified or distorted, and the use of the
computer in pursuing higher level teaching objectives such
as problem-solving.

Although this early work shows promise and has produced
many stimulating ideas, there is a need for a substantial
coherent programme of research and curriculum development
in mathematical education, with a conscious awareness of the
possibilities of the new technologies. Proposals will no
doubt be made under the Government's *Microelectronics
Development Programme for Schools and Colleges*; they will
include a proposal based on the two existing projects already
mentioned and the centres for research in mathematical education at Nottingham and Chelsea. They must aim both to
provide help to teachers in teaching the current curriculum
and an exploration of how that curriculum may be modified
over the next decade or so to reflect the presence of cheap
computing power in society and in the classroom.

References for Section 1, 'Calculators in the Classroom'

AYERS, S. (1977): 'The effects of situational problem-solving
and electronic calculating instruments in a college level
introductory statistics course. Georgia State University,
1976', *Diss Abs Int* 37A, 6322-3.

BELL, A.W., BURKHARDT, H., McINTOSH, A. and MOORE, G. (1979):
A Calculator Experiment in a Primary School. Shell Centre
for Mathematical Education, University of Nottingham.

CONNAH, P. (1979): The effect on candidates' scores of allowing the use of electronic calculators in examinations at 16+. B.Phil.Ed., University of Birmingham.

COUNTY OF DURHAM EDUCATION COMMITTEE (1976): *The Use of Electronic Calculators in Secondary Schools*. Darlington Teachers' Centre.

HOPKINS, W.L. (1978): The effect of a hand-held calculator curriculum in selected fundamentals of mathematics classes. Unpublished Doctoral Dissertation, University of Texas.

KASNIC, N.J. (1978): 'The effect of using hand-held calculators on mathematics problem-solving ability among sixth grade students. Oklahoma State University, 1977', *Diss Abs Int*, 38A, March, 5311.

LOWERRE, G.F., VENESKI, J., SCANDURA, A.M. and SCANDURA, J.M. (1976): 'Using electronic calculators with elementary school children', *Ed Tech*, 16, 14-8.

OXFORD AND CAMBRIDGE SCHOOLS EXAMINATION BOARD (1979): Awarders' report: Awarders' comments on the SMP O-level Calculators Paper, Summer.

SMP (1977): *Calculators in Schools: Report of the investigation 1976-77*. SMP, Westfield College, Kidderpore Avenue, Hampstead, London NW3 7ST.

SMP (1979): *SMP Calculator Series: Calculator Supplement by Topics, Calculator Supplement to Books X, Y and Z, Discover How to Use Your Electronic Calculator, Growth and Decay: Financial and Other Applications, Sequences and Iterative Processes*. Cambridge University Press.

SUYDAM, M.N. (1979): *The Use of Calculators in Pre-College Education: A State-of-the-Art Review*. Calculator Information Center Newsletter, May 1979. Calculator Information Center, 1200 Chambers Road, Columbus, Ohio 43212.

SUYDAM, M.N. (1980): *References on Instructional Activities, Research Reports, and Other Topics Related to Calculator Use*. Calculator Information Center Newsletter, May 1980, Reference Bulletin No. 25.

ZEPP, R.A. (1976): 'Reasoning patterns and computation on proportions problems, and their interaction with the use of pocket calculators in ninth grade and college. Ohio State University, 1975', *Diss Abs Int*, 38A, February, 5181.

References for Section 2, 'Computers in Mathematics Education'

Complete references for research reported in this section are available from Professor D.C. Johnson, Centre for Science Education, Chelsea College, Bridges Place, London.

Chapter Fourteen
Using Research

1.   INTRODUCTION

We shall indicate in this chapter some of the main sources and indexes through which British and American research can be traced. Mention should first be made of a chapter contributed by A.G. Howson to *The Use of Mathematical Literature* (Ed. A. Dorling, 1977), which includes much of the material described here, and also a range of source books and addresses of institutions and associations, including some from Europe as well as from Britain and the United States.

2.   BRITISH RESEARCH

Journals and reports

Reports of British research in mathematics education appear in a variety of publications. Some work, along with reports from other countries, is published in *Educational Studies in Mathematics*: this is useful because it contains not only individual studies but reviews of whole areas (e.g., Austin and Howson, 1979, on language; Fennema, 1979, on sex-differences). A certain amount of psychologically-based research can be found in the *British Journal of Educational Psychology*: many articles are concerned specifically with mathematics, but, even amongst those which are apparently not so specific, results relevant to mathematics education

can be identified. Some research is described, usually in a
broad, descriptive style in *Mathematics Teaching* or
Mathematics in School, or even occasionally (though recently
less so) in the *Mathematical Gazette;* and the *International
Journal of Mathematics Education for Science and Technology*
carries some articles relevant mainly to teaching in
universities and polytechnics. From time to time, reports of
relevant research appear in *Mathematical Education for
Teaching, Cambridge Journal of Education, Forum, Educational
Research, Journal of Curriculum Studies, Educational Studies,
Where* and even the *Times Educational Supplement*.

Some institutions publish and distribute their own
research reports: lists of such material are available from
Brunel University School of Education, the Computer-Based
Learning Unit at the University of Leeds Centre for Studies
in Science Education, the Artificial Intelligence Department
at the University of Edinburgh, the Shell Centre for
Mathematical Education at the University of Nottingham. The
Centre for Science Education at Chelsea College, London,
publishes the termly lectures given to its Psychology and
Mathematics Education Workshop.

Theses and dissertations

UK theses and dissertations concerned with mathematics
education presented for higher degrees in the United Kingdom
are now being listed by Frobisher and Joy of Leeds Polytechnic, in

*Mathematical Education, a Bibliography of Theses and
Dissertations* (from Mathsed Press, Fairfield House,
68c Gledhow Wood Road, Leeds, LS8 4DH).

Again, this is a very useful reference work: the theses
are listed by year of presentation. Titles are given in full;
abstracts are not given, but there is a keyword index; and
within its restricted framework the listing is comprehensive.
Supplements are available annually.

Current work is listed in the JMC/Shell Centre document
Current Research and Development in Mathematical Education.
This is broadly categorized and contains also a list of
correspondents.

Indexes and other source material

As far as British journals are concerned, a valuable source
is provided by

References of Use to Teachers of Mathematics

published in 1974 by the Centre for Studies in Science
Education at Leeds University, together with the Supplement

No.1, published in 1978. This pulication is an index, organized under subject headings, to the following journals:

International Journal of Mathematical Education in Science and Technology
The Mathematical Gazette
Mathematics in School
Mathematics Teaching
Mathematical Education for Teaching
Education for Teaching
The Bulletin of the Institute of Mathematics and its Applications
Education in Chemistry
Education in Science
Physics Education
Journal of Biological Education
School Science Review
Educational Studies in Mathematics (only the articles in English)
British Journal of In-service Education
Educational Research

References go back to 1961, but some journals are included only in the supplement, so that both sections are needed to give coverage of the period 1961-78.

The Leeds listing does not provide an exhaustive coverage of journal articles on mathematics education; a sample of a few pages of *British Education Index* indicates that about 30 per cent of the articles listed under 'mathematics' are from other sources. *British Education Index* cannot of course provide the specialized subject headings of the Leeds publication, but it does pick up articles from a wider range of journals; relevant articles appear in some of these only rarely, but in others, particularly the *British Journal of Educational Psychology* and the *Times Educational Supplement*, fairly regularly. *References of Use to Teachers of Mathematics* is just what the title implies, and in its choice of journals adopts a middle road between the popular feature article (as in T.E.S.) and the theoretically based research study (as in B.J.E.P.).

Neither of these bibliographies gives more than a title, so that it is not always completely clear what the article is about, and of course theses, books and various other sources are not included.

3. AMERICAN RESEARCH

Journals and reports

The main American research journal is the *Journal for Research in Mathematics Education*. Other articles appear in the *Journal of Children's Mathematical Behaviour, Arithmetic*

322

Teacher, Mathematics Teacher, and in *School Science and Mathematics;* and review articles relating to aspects of mathematics education are published in *Review of Educational Research.*

The ERIC Clearing House for Science, Mathematics and Environmental Education, at 1200 Chambers Road, Columbus, Ohio, publishes substantial reports of conferences and surveys of fields of research, as well as the essential bibliographic material. A list of currently available material is available. We have found these to be essential sources for this document.

Indexes and sources

ERIC publishes the following:

1. *Mathematics Education Abstracts and Index from RIE:*

	1966-72
(ERIC documents: mainly research	1973-75
reports and teaching units)	1976-77

2. *A Categorised Listing of Research*   1966-73
                                           1974-78

   (Articles, dissertations, ERIC documents)

3. *Investigations in Mathematics Education*

   (Critical Abstracts of selected articles and listings of just-published reports)

Annual un-categorized listings of articles and dissertations appear in the July issue of the *Journal for Research in Mathematical Education.* Abstracts of all dissertations appear in *Dissertation Abstracts International,* and University libraries can usually obtain through the British Library reprints of these dissertations. ERIC documents are also available through the British Library.

Supplement

This supplement is an attempt to update, where necessary, the comments and conclusions of the review in the light of more recent work. The sources quoted here are easily accessible, and often include review articles with substantial bibliographies, so that those interested can follow up those areas which particularly concern them. Note also the publications referred to in the supplement to Chapter 14, in particular the annual proceedings of the PME.

CHAPTER 3:   STAGES OF GENERAL INTELLECTUAL DEVELOPMENT

The extent to which general stages of cognitive development
are demonstrated in the Chelsea CSMS project has been
discussed by Küchemann (1981).   Two contrasting views are
recognized of how children's understanding develops.   One
view is that development is highly individualistic and can be
markedly influenced by what is taught: another is that understanding develops in a sufficiently uniform way for children
usefully to be assigned to stages of cognitive development
and that age-related constraints on this development limit
the effects of teaching.   Küchemann does not try to resolve
these conflicting views, but rather adopts the latter viewpoint as a framework for examining the results of the CSMS
mathematics tests.   The discussion focusses on the mismatch
between children's understanding of mathematics and the
cognitive demand of much of what they are taught.   Some
general dimensions are identified which could provide useful
guidelines to teachers about the nature and order of magnitude of the factors which affect the difficulty of mathematical tasks; and the extent to which it is possible to
identify common features between different items at a given
facility level is impressive.

In a recent review article, Hiebert and Carpenter (1982)
have considered the usefulness of Piagetian tasks as
readiness measures in mathematics teaching.   The research
consistently indicates that, while there is a positive
correlation between performance on Piagetian tasks and
mathematics achievement, many school mathematics tasks can be
completed successfully by children who have apparently not
yet developed the relevant reasoning ability measured by
Piagetian tests.   This, however, is not a criticism of
Piagetian theory.   It simply draws attention once again to
the different aspects of competence involved in carrying out
a specific learnt process on the one hand, and in understanding and recognizing the underlying mathematical
structure on the other.   The fact that performance on most
school tasks does not depend on the presence of associated
Piagetian abilities is a reflection of the kind of mathematics which children are usually expected to master.

CHAPTER 4:   SKILLS, CONCEPTS and STRATEGIES

The distinctions between the various components of mathematical competence, while still in some respects crudely
defined, are increasingly recognized.   One of the more
interesting comparisons of attainment in computation and
recognition of operation is provided by Matthews (1981).   In
her study, the written computation task

    394 - 197 = ...

shows a steady facility of about 70 per cent amongst

11-year-olds and various groups of adults, whereas the facility of 'story' tasks involving recognition of subtraction is improved by age and experience. One example shows an increase from 18 per cent success rate with 18-year-olds to about 60 per cent with adults. This finding is consistent with the differences reported by NAEP in the United States.

CHAPTER 5: THE TEACHING OF FACTS AND SKILLS

There is a section by Suydam and Dessart in the NCTM review of research (Shumway (Ed), 1980), entitled 'Skill Learning' which summarizes work on the development of mathematical skills. This adds a considerable amount to our Chapter 5.

The part of this chapter, however, which has seen the most extensive development during recent years is section 5, concerned with the contrast between standard algorithms and children's methods. This contrast has been recognized in many studies, some concerned simply with analysing children's understanding and others with developing effective teaching approaches. Streefland's work (1982) on teaching the subtraction of fractions, for example, is intended to respect the points of view of the children, meeting the children's need to build their own algorithms. There are a number of illustrations of children's methods in the programmes associated with the Open University course EM 235, Developing Mathematical Thinking. Many children are shown subtracting three-figure numbers successfully in an intuitive way, while being quite unable to use the standard algorithm. See also the book edited by Floyd (1981).

There is an increasing awareness that while children may develop their own methods rather than rely on taught algorithms, many continue to use informal, 'naive' methods which are limited in their applicability rather than adopt the standard, systematic procedures supposedly developed in secondary school. The suggestion that children hold official mathematics and commonsense calculation in separate compartments is strongly supported by a number of examples in the CSMS tests, as illustrated by Booth (1981). 'Child methods', restricted mainly to counting, adding-on or building-up are a common source of errors (see Hart, 1981a). Indeed Booth suggests that the difficulties which some children experience in mathematics are due to these children's non-initiation into the formal, taught system. If a child is in fact operating within a system of his own which is different from formal mathematics, then

a) merely demonstrating methods in the formal system will have little success;
b) restricting the methods taught to those which are consonant with the child's will be ultimately ineffective, so long as these methods remain defined by the child's

approach,

c) ways must be found of making the child aware of the limitations of his own approach and of the existence and power of the formal system.

Considerations such as these are the background to the new Chelsea College project, Strategies and Errors in Secondary Mathematics. The task of this project is to determine the validity of notions like a), b) and c) above, and, where appropriate to develop ways of translating these ideas into effective classroom practice. Some indication of the kind of work involved in this project is given by Booth (1982).

The comments in the final paragraph of Chapter 5 about Skemp's distinction between instrumental and relational understanding are illustrated and further developed in a different topic area by Pollatzele *et al.* (1981) in a discussion of students' understanding of the statistical idea of the mean. This work involves a number of different perspectives. As a consideration of students' understanding of statistics, it relates to part of Chapter 6; and its consideration of the effects of context and presentation is relevant to the discussion of word problems in another part of that chapter.

CHAPTER 6: CHILDREN'S UNDERSTANDING OF CONCEPTS IN SPECIFIC TOPIC AREAS

As indicated previously, there are a number of articles which involve several different aspects of research. This is especially true in relation to Chapter 6, where some articles include a consideration of the difference between learning skills and concepts within a particular topic while others combine the analysis of children's understanding of a topic with a teaching experiment and suggestions for the classroom.

However, one study which concentrates on children's understanding of a particular concept is that by Bednarz and Janvier (1982) on the understanding of the numeration system. The objectives of this research are: to clarify the notion of numeration, to clarify what is meant by under-standing numeration, to develop a reference framework to evaluate understanding, and to single out children's difficulties and strategies. Difficulties in handling groupings, especially in 'un-making' groupings, and in using zero, are particularly noted.

The recognition of numerical operations in verbal problems has received increasing attention. Nesher *et al.* (1982) interpret performance in terms of variables of a linguistic nature. Ballew and Cunningham (1982) identify a number of factors which affect the difficulty of word problems and describe a diagnostic system for determining individual difficulties, particularly involving reading, problem-interpretation and problem-solving. Hubert (1982) suggests that problems which are amenable to direct representation may be easier to solve than those that are not. This is a development of work by Carpenter *et al.* (1981) which shows that many young children are able to solve verbal addition and subtraction problems by representing the sets in the problems with objects, such as counters or cubes, and carrying out the prescribed action on the objects. Bell, Swan and Taylor (1981) discuss the factors affecting the difficulty in recognizing arithmetical operations and then describe a diagnostic procedure, and a teaching experiment which attempts to improve pupils' ability to choose the correct operation. Three strategies are developed: drawing diagrams, using easier numbers and estimating. Considerable success was achieved in improving understanding of place value, with a more modest improvement in correct choice of operation.

The CSMS work on fractions which is given a brief mention in Chapter 6 is discussed and summarized by Hart (1981b). Two articles on the understanding of fractions by Hasemann (1981) and Peck and Jencks (1981) discuss the performance of children in Germany and the United States respectively, and have remarkably similar conclusions. In both cases, children are observed to go through the motions of operations with fractions but are not able to attach any meaning to these processes. This is a familiar theme, which appears to be particularly starkly illustrated in the context of operations with fractions.

The topic of directed number is a favourite area for teaching experiments involving 'embodiments'. Some examples of these were described in our Chapter 7, section 7. Two articles in similar vein have appeared recently (Arcavi and Bruckheimer 1981, and Rowlands 1982): these are concerned both with children's understanding of the topic and with suggestions for situations which might provide an effective teaching approach.

The comments made under the topic heading 'Algebra' about children's interpretation of the equals sign are reinforced in a study by Kieran (1981). The notion that the equals sign is a 'do something' signal, is characteristic of many young children's thinking, and is shown to remain, albeit in more subtle forms, amongst older children.

A comprehensive review of research related to the 'topological primacy thesis' is provided by Darke (1982). The article includes a critique of many aspects of Piaget's assertion that the order of acquisition of geometrical ideas is first topological concepts, then projective, then euclidean, and concludes that it is ill-supported and generally misapplied.

There has been some recent work on Estimation, a topic
not specifically considered in Chapter 6. Reys *et al*. (1982)
discuss computational estimation and identify the processes
used by good estimators. Siegel *et al*. (1982) focus on
estimation of measurements, and provide a tentative model of
estimation applicable to both number and measurement. Levin
(1981) discusses a range of estimation techniques, and
presents in detail a series of mental estimation procedures.
The article includes a description and some evaluation of
techniques for teaching these procedures.

CHAPTER 7:   ASPECTS OF TEACHING METHOD

The teaching of conceptual structures is a particular concern
of Chapter 7, and section 7 on embodiments considers some
situations through which abstract notions may be constructed
and recognized. A study by Janvier (1981) discusses the use
of complex 'realistic' situations, and raises various issues
regarding their use on the basis of the responses of children.
The role of mental images and the importance of spoken
language are both emphasized in this work, and the need for
the gradual drawing out of the salient features is discussed.

CHAPTER 8:   PROCESS ASPECTS OF MATHEMATICS

A review of research on mathematical problem-solving is
provided by Lester in Chapter 10 of Shumway (Ed), (1980).

The range of different problem-solving heuristics which
may usefully be employed in a particular situation is
illustrated by Kraus (1982) in an investigation of how people
play Nim. Strategies include deduction by synthesis (working
forward), analysis (working backward), trial and error,
looking for patterns, reviewing previous work, using a related
problem, using pictorial representations and using subgoals.
Moreover, Kraus' use of cluster analysis divides players into
three categories according to the kinds of strategies they
are inclined to adopt. It is suggested that research should
be continued with a view to the effective incorporation of
the use of games in the teaching of problem-solving.

Section 3 of our Chapter 8 is concerned with 'real
problem' solving; this is the subject of an article by Lesh
(1981). Lesh reviews a substantial amount of earlier work
and presents a case for focussing research on

(1)   average ability students;
(2)   substantive mathematical content;
(3)   real problems; and
(4)   realistic settings and solution procedures.

It is noted that restrictions on solution procedures may turn
'real' situations into artificial ones: good problem solvers
work not in isolation but through the effective use of

resources - technological tools (e.g. calculators), other
people (e.g. classmates), and other resources (books,
newspapers etc.).

Work which carries forward that reported in the section
on Proof has been done by Galbraith (1981). He is concerned
with the perceptions which secondary school children have of
forms of mathematical argument. Modes of logical deduction
have an agreed status within mathematics, but differences
exist between the meaning and purpose of some mathematical
terms and procedures and the way in which these are used by
students. By considering levels of performance, it is
possible to identify particular components which consistently
determine the capacity to construct or follow proofs and
explanations.

In our consideration of deductive geometry and
Van Hiele's levels, we noted that the trend in Britain is
against the use of geometry as a means of developing the
appreciation of deductive argument and the need for proof.
Lang and Ruane (1981) have documented this development in
England; they suggest a way to build on an experimental
approach to geometry in order to provide a unified knowledge
of spatial configurations through deductive methods.

CHAPTER 9: ATTITUDES

Pupils' attitudes to mathematics are considered in greater
depth and from a somewhat different standpoint by Hoyles
(1982). She reports an exploratory study which examined how
14-year-old pupils perceive good and bad learning experiences
in school. In particular, some significant features
associated with learning mathematics are described. The
study suggests that pupils have strong ideas about what they
are capable of doing and understanding in mathematics, and
that their mathematical experience is dominated by this focus
on themselves and their feelings. However, the diversity of
the stories which are quoted suggests that pupils differ
considerably in the goals they set themselves. Some like
being able to do mathematics on their own, some enjoy
challenge, and some are well-satisfied if they are told what
to do to reach the solution; many are concerned with
obtaining the right answers and with the marks received.
Despite these differences of goals, failure appears to create
similar doubts and feelings of inadequacy. The following
quotation illustrates the tendency:

'I just wanted to get something down on paper, that's all
... just be able to write down a few lines to show I'd at
least tried and was not completely stupid. It was no good.
I was just a failure ... I knew I would never be able to
get anywhere with it, no matter how long I sat there ...'

Some further evidence about the relationship between attitude and achievement in mathematics is provided by Matthews (1981). The positive (but relatively low) correlation is again confirmed. A group of nursery nurses showed an attitude of stated dislike, at times expressed violently, towards mathematics, and yet were able to achieve at least comparable success with other occupational groups. It is suggested that the nursery nurses were 'role-playing', possibly subconsciously, as being 'no good at maths'.

CHAPTER 10: INDIVIDUAL DIFFERENCES

The relationship between individual differences and the learning of mathematics is discussed by Fennema and Behr in Chapter 11 of Shumway (1980). The mere identification of traits on which individuals differ is not considered by the authors to be a particularly profitable area for research, unless the implications of these differences for learning and teaching mathematics are clearly ascertained.

Some sections of our Chapter 10 on individual differences have been further developed by Head (1981). He draws attention to the lack of understanding of the link between personality variables and performance in mathematics, even though it is commonly (and reasonably) assumed that these variables play an important role, especially in issues such as motivation, attitude and cognitive style.

The part of this work which has seen most development in recent years is the study of gender-related differences in learning mathematics. There is a review of research in this area by Badger (1981). Appendix 2 of the Cockcroft Report (1982) is a discussion by Shuard of differences in mathematical performance between girls and boys, and includes some suggested strategies for improving the balance. Fennema (1981) has described a study of intervention, designed to increase women's participation in mathematics: her results suggest that such intervention can be effective in increasing female students' enrolments in mathematics courses.

CHAPTER 11: LANGUAGE

Further discussion of the use of precise symbolism and linguistic structure in teaching mathematics (Chapter 11, section 7), is provided by MacKerman (1982). He draws attention to some instances where, contrary to the norm, it might be advisable to use words rather than symbols in written mathematical expressions.

There are a number of recent studies which identify some particular difficulties experienced by students and children who are either bilingual or are learning mathematics in a weakly known, second language. These disadvantages have so far had little recognition: the learning of mathematics has appeared to be affected less than most other subject areas.

Zepp (1982), in a study of children in Lesotho secondary
schools, considers particularly connectives such as 'and',
'or' and 'if ... then'. Some incorrect reasoning patterns
are displayed, attributable to linguistic differences and
difficulties. Jones (1982) has investigated the understanding of the terms 'more' and 'less' by children in
Papua New Guinea. His work shows how delay in acquisition of
meanings of such terms can lead to a mismatch between the
demands of the curriculum and the language competence of the
learner. These studies are concerned with, and show the need
for investigation of, the specific features of particular
languages which impinge directly on the learning of mathematics.

CHAPTER 12: EVALUATION AND ASSESSMENT

The APU has now completed its series of annual surveys of the
mathematical performance of 11- and 15-year-olds which began
in 1978. Testing in mathematics will not continue annually,
but there are plans to monitor performance at five-yearly
intervals. Six reports of 'Mathematical Development' have
been published (1980-82); and Foxman (1982) has provided
an outline of some of the main conclusions to be drawn after
five years of testing. For example, he suggests that the
APU's individual tests of practical mathematics represent a
significant development in assessment. The procedures for
administering these tests have been sufficiently refined and
standardized for practical testing to be a serious
proposition as a component of public examinations.

The APU publications have received a certain amount of
criticism (see, for example, Preston 1980), often on the
grounds that the results are published without any analysis
or interpretative commentary. Thus a statement such as
'34 per cent of the pupils gave the correct answer to "what
is the cube of 4?"', is at best unhelpful and perhaps
meaningless. The results of an individual item cannot
indicate why pupils succeeded or not, but information on this
may sometimes be gleaned from similar or contrasting items.
Thus, one example from the third secondary report demon-
strates how, in items on square roots, the facility and the
nature of errors are affected by the different conventions
used to indicate a square root, and by the size of the
numbers involved. In the later surveys, a more extensive
analysis of errors has been undertaken, and this reveals
many cases where the incidence of particular errors varies
considerably between questions with apparently similar
structure. The factors involved are gradually becoming more
clearly recognized. It is clear that the extent of a pupil's
grasp of a particular concept cannot be revealed by a
limited range of examples.

CHAPTER 13: CALCULATORS AND COMPUTERS

Research into the effectiveness of calculators in two
particular topic areas is described in recent articles by
Behr and Wheeler (1981), concerned with early number concepts
and counting, and by Szetela (1982) on identifying and
carrying out the appropriate arithmetical operations in word
problems.

In respect of calculators and computers in mathematics
education, Chapter 7 of the Cockcroft Report (1982) notes
particularly the weight of evidence that the use of calculators has not produced any adverse effect on computational
ability.

CHAPTER 14: USING RESEARCH

The book *Research in Mathematics Education,* referred to in
several parts of this supplement, was published by NCTM in
1980.   Edited by Richard Shumway, it is an effort to produce
a definitive reference work on research in mathematics
education.   The first part of the book - something like one
third of the text - is concerned with research process and
methodology.   This is followed by eight chapters dealing with
research in particular aspects of mathematics education.
Many of the chapters are comparable in scope to ours; but the
references quoted are almost exclusively from the United
States.

Probably the best sources of information about current
research are the proceedings of the annual conferences of the
International Group for the Psychology of Mathematics
Education (PME).   These are distributed in Europe by the
Shell Centre for Mathematical Education at the University of
Nottingham (current price £6 per volume), and in North
America by the Wisconsin Research and Development Centre for
Individualized Schooling, Madison, Wisconsin.

References

ARCAVI, A. and BRUCKHEIMER, M. (1981): 'How shall we teach
the multiplication of negative numbers?' *Maths in School,*
10, 5.

ASSESSMENT OF PERFORMANCE UNIT (APU) (1980-82): *Mathematical
Development; Primary Survey Nos 1 to 3 and Secondary Survey
Nos 1 to 3.*   London: HMSO.

BADGER, M.E. (1981): 'Why aren't girls better at maths? A
review of research', *Ed Res,* 24, 1.

BALLEW, H. and CUNNINGHAM, J.W. (1982): 'Diagnosing Strengths
and weaknesses of Sixth-Grade Students in Solving Word
Problems', *J Res Math Ed,* 13, 3.

BEDNARZ, N. and JANVIER, B. (1982): 'The understanding of numeration in primary school', *Ed Stud Math*, 13,1.

BEHR, M.J. and WHEELER, M.M. (1981): 'The calculator for concept formation: a clinical status study', *J Res Math Ed*, 12, 5, 323-38.

BELL, A.W., SWAN, M. and TAYLOR, G. (1981): 'Choice of operation in verbal problems with decimal numbers', *Ed Stud Math*, 399-420.

BOOTH, L.R. (1981): 'Child-methods in secondary mathematics', *Ed Stud Math*, 12,1.

BOOTH, L.R. (1982): 'Ordering your operations', *Maths in School*, 11,3.

CARPENTER, T.P. *et al.*(1981): 'Problem structure and first-grade children's initial solution processes for simple addition and subtraction problems', *J Res Math Ed*,12, 1, 27-39.

COCKCROFT, W.H. (1982): *Mathematics Counts*. London: HMSO.

DARKE, I. (1982): 'A review of research related to the topological primacy thesis', *Ed Stud Math*, 13, 2.

FENNEMA, E. *et al.* (1981): 'Increasing women's participation in mathematics: an intervention study', *J Res Math Ed*, 12, 5.

FLOYD, A. (Ed) (1981): *Developing Mathematical Thinking*. Open University/Addison-Wesley.

FOXMAN, D. (1982): 'Testing, testing, one, two, three ...' *Times Ed Supp*, 10 Dec 1982.

GALBRAITH, P.L. (1981): 'Aspects of proving: a clinical investigation of process', *Ed Stud Math*, 12, 1.

HART, K. (1981a): 'Investigating Understanding', *Times Ed Supp* 27 March 1981.

HART, K. (1981b): 'Fractions', *Maths in School*, 10, 2.

HASEMANN, K. (1981): 'On difficulties with fractions', *Ed Stud Math*, 12, 1.

HEAD, J. (1981): 'Personality and the learning of mathematics', *Ed Stud Math*, 12, 3.

HEIBERT, J. (1982): 'The position of the unknown set and children's solutions of verbal arithmetic problems', *J Res Math Ed*, 13, 5.

HIEBERT, J. and CARPENTER, T.P. (1982): 'Piagetian tasks as readiness measures in mathematics instruction: a critical review', *Ed Stud Math*, 13, 3.

HOYLES, C. (1982): 'The pupil's view of mathematics learning', *Ed Stud Math*, 13,4.

JANVIER, C. (1981): 'Use of situations in mathematics education', *Ed Stud Math*, 12, 1.

JONES, P.L. (1982): 'Learning mathematics in a second language: a problem with more or less', *Ed Stud Math*, 13, 3.

KIERAN, C. (1981): 'Concepts associated with the equality symbol', *Ed Stud Math*, 12, 3.

KRAUS, W.H. (1982): 'The use of problem-solving heuristics in the playing of games involving mathematics', *J Res Math Ed*, 13, 3.

KÜCHEMANN, D.E. (1981): 'Cognitive demand of secondary school mathematics items', *Ed Stud Math*, 12, 3.

LANG, B. and RUANE, P. (1981): 'Geometry in English secondary schools', *Ed Stud Math*, 12, 1.

LESH, R. (1981): "Applied mathematical problem solving', *Ed Stud Math*, 12, 2.

LEVIN, J.A. (1981): 'Estimation techniques for arithmetic: everyday math and mathematics instruction', *Ed Stud Math*, 12, 4.

MACKERNAN, J. (1982): 'The merits of verbalism', *Maths in School*, 11, 4.

MATTHEWS, J. (1981): 'An investigation into subtraction', *Ed Stud Math*, 12, 3.

NESHER, P. *et al.* (1982): 'The development of semantic categories for addition and subtraction', *Ed Stud Math*, 13, 4.

PECK, D.M. and JENCKS, S.M. (1981): 'Conceptual issues in the teaching and learning of fractions', *J Res Math Ed*, 12, 5, 339-48.

POLLATSEK, A. *et al.* (1981): 'Concept or computation: students' understanding of the mean', *Ed Stud Math*, 12, 2.

PRESTON, M (1980): 'The first APU mathematics survey', *Education 3-13*, Autumn.

REYS, R.E. *et al.* (1982): 'Processes used by good computational estimators", *J Res Math Ed*, 13, 3.

ROWLAND, T. (1982): 'Teaching directed numbers: an experiment', *Maths in School,* 11, 1.

SHUMWAY, R.J. (Ed) (1980): *Research in Mathematics Education.* NCTM, Reston, Virginia.

SIEGEL, A.W. *et al.* (1982): 'Skill in estimation problems of extent and numerosity', *J Res Math Ed,* 13, 3.

STREEFLAND, L. (1982): 'Subtracting fractions with different denominators', *Ed Stud Math,* 13, 3.

SZETELA, W. (1982) 'Story problem solving in elementary school mathematics: what difference do calculators make?', *J Res Math Ed,* 13, 5.

ZEPP, R.A. (1982): 'Bilinguals' understanding of logical connectives in English and Sesotho', *Ed Stud Math,* 13, 2.